Cosmos and the Rhetoric of Popular Science

Cosmos and the Rhetoric of Popular Science

Karen Schroeder Sorensen

LEXINGTON BOOKS
Lanham • Boulder • New York • London

Published by Lexington Books
An imprint of The Rowman & Littlefield Publishing Group, Inc.
4501 Forbes Boulevard, Suite 200, Lanham, Maryland 20706
www.rowman.com

Unit A, Whitacre Mews, 26-34 Stannary Street, London SE11 4AB

British Library Cataloguing in Publication Information

Library of Congress Cataloging-in-Publication Data Available
Names: Sorensen, Karen Schroeder, author.
Title: Cosmos and the rhetoric of popular science / Karen Schroeder Sorensen.
Other titles: Cosmos (Television program)
Description: Lanham : Lexington Books, [2016] | Includes bibliographical references and index.
Identifiers: LCCN 2017021051 (print) | LCCN 2017024190 (ebook) | ISBN 9781498507608 (Electronic) | ISBN 9781498507592 (cloth : alk. paper)
Subjects: LCSH: Astronomy. | Cosmology. | Television in astronomy.
Classification: LCC QB43.3 (ebook) | LCC QB43.3 .S628 2016 (print) | DDC 520.1/4--dc23
LC record available at https://lccn.loc.gov/2017021051

∞™The paper used in this publication meets the minimum requirements of American National Standard for Information Sciences—Permanence of Paper for Printed Library Materials, ANSI/NISO Z39.48-1992.

Printed in the United States of America

Contents

This book is dedicated first to my love, Alan Sorensen. Words won't come close to doing you justice, so I'll keep it short: without you, our lives would be tremendously different. Thank you.

It's also dedicated to my daughters, Emmelyn Jane and Jillaine Carole. You know that I persisted. And you will also persist. You are strong women who will create your own lives and be happy on your own terms. I'm so lucky to get to be your mom.

Acknowledgments

I would like to acknowledge the help of the countless people along the way who challenged me and helped me on my "first-gen" journey. I'll name a few who are particularly important to me below.

Dr. Dale Sullivan, for providing me with a graduate teaching fellowship that included health insurance. Without it, I couldn't have gone to graduate school. He's been a mentor and a friend. A genuinely good man.

Dr. Elizabeth Birmingham, for her encouragement throughout my academic and professional journey. She's the kind of teacher I hope to be.

Dr. Thomas Matchie and Dr. Muriel Brown, for supporting me during difficult times long ago and letting me know that I had potential.

Dr. April Herndon and Dr. Henry Gonshak, for providing me with words of wisdom (and comfort) about publishing and peer review as I went through that process.

Paula McGarvey, for offering feedback and perspective as I drafted.

Wesley and Carole King, because I like them and they deserve it.

Erin Kreps, my sister in life, for being my friend. I am so lucky to know her.

Kelly and Scott Seago, for being the best cheerleaders on this planet (and maybe any other).

Bessie Babcock Deyo, for letting me know that I could accomplish something with my life. Her words still ring in my ears and they continue to shape me. Grandma, I miss you.

Larry Schroeder, for teaching me to think about situations logically and empathetically. He was a man who deserved more than he got and I think I am benefiting from his good karma. Dad, I miss you.

Winona State University, for being a school that truly cares about its students, faculty, and staff. The English department especially has created an academic environment that is collegial, professional, and supportive. Because of this positive environment, I was able to (finally) finish this book.

Chapter 1

Cosmos' Epic Journey

Cosmos: A Personal Voyage, an educational miniseries that premiered on PBS in September of 1980, embodies the rhetorical promise of popular science television. Hosted by Carl Sagan, a scientist and one of the most important science popularizers in history, the program became a popular science juggernaut, re-airing many times. It brought Hollywood-level special effects to televisions and living rooms across the nation when many homes still didn't have cable. It was reminiscent of popular entertainment of the day and was often compared to *Star Wars*. Concepts of evolution, space travel, and extraterrestrial life were presented to the public in a way that engaged viewers and entertained them. In doing so, Sagan became a model among the scientific popularizers of the late twentieth century.[1]

An educational program—particularly one made up of only thirteen episodes—rarely has the lasting impact of *Cosmos*, a series that has affected both popular culture and the public perception of science for over thirty years. Sagan's opus takes viewers on a journey across the universe and through its history. It shows the formation of stars and the history of scientific advances on the Earth. Through it, viewers visit ancient Alexandria and "see" the long-destroyed library there that once held "all the knowledge of the ancient world."[2] Throughout the program, Sagan repeatedly warns viewers of the dangers of nuclear weapons and encourages them to acknowledge the value and uniqueness of human life—even while simultaneously illustrating that Earth is but one tiny speck within an unimaginably massive universe. To its viewers, *Cosmos* became more than an educational program: it became a cultural moment, a narrative of our civilization and of the universe.

Cosmos was popular from the beginning, and according to biographer Keay Davidson, it was the "climax of [Carl Sagan's] ascent to fame."[3] Sagan had already published a number of books before *Cosmos*; one had even been awarded the Pulitzer Prize.[4] He was also the regular "house astronomer" on *The Tonight Show with Johnny Carson.*[5] Clearly, Sagan was not a complete unknown before *Cosmos*. However, he did not achieve "iconic status in North American culture,"[6] which made his name and his face recognizable in households across the nation, until after this incredibly popular program premiered. In the year-2000 reissue of *Cosmos* on DVD, series cocreator and cowriter Ann Druyan notes: "the series catapulted creator and host Carl Sagan to the status of pop culture icon and opened countless minds to the power of science and the possibility of life on other worlds."[7]

Few could have predicted *Cosmos*' durability over the next three decades, even though, as Thomas Lessl reports, its "budget of eight million dollars [made it] the most expensive program in the history of public television"[8] up to that time. In fact, *Cosmos* still has enough name recognition cachet that instead of creating a new science series from scratch, FOX did a soft reboot of the brand in 2014. Seth MacFarlane, of *Family Guy* fame, produced it and Neil deGrasse Tyson hosted.

In some ways it's not surprising that the program has been resurrected. The early twenty-first century seems focused on (even obsessed by) program and movie reboots. These modernized takes on long-loved shows include *Batman*, *The Avengers*, *Star Wars*, and *Star Trek*. However, educational programs aren't normally part of that picture. *Mr. Rogers' Neighborhood* and Bob Ross's *The Joy of Painting*, for example, remain untouched despite their notoriety. What makes these programs different? How has *Cosmos* managed to remain revered as an iconic pop culture moment? What makes the program so appealing? This book seeks to provide potential answers to these questions by expanding on Thomas M. Lessl's original insights[9] regarding the program. It will explore what rhetorical elements contribute to *Cosmos'* creation of *mythos* (via fantasy-theme analysis) and examine *Cosmos* in terms of the rhetorical concepts of *kairos* and *ethos*.

Just as Fred Rogers's and Bob Ross's identities in the public realm are inextricably intertwined with their programs, so is Sagan's name synonymous with *Cosmos*. His programmatic persona is that of a favorite teacher, filling the gaps of knowledge and opening the audience's eyes to new concepts. At the same time, the science he delivers contains hard truths, criticisms, and even accusations. Yet, despite his potentially divisive challenges in regard to war, politics, and evolution, *Cosmos* became the most successful PBS series in history—and it remained so for a decade.[10] Sagan's rhetorical approach helped to make him, and his signature program, immensely popular; through a "well-wrought phrase or gripping metaphor,"[11] he captured the attention and imagination of his audience.

The ways in which *Cosmos* intersects popular science with science fiction provide fertile ground for rhetorical investigation. Discovering what rhetorical frameworks Sagan used in *Cosmos* may help to construct a sort of pattern that could be applied to other popular science artifacts—particularly those on television. This rhetorical investigation intends to provide insight into the ways in which *Cosmos* appealed to its contemporary audience and became a landmark work of popular science. It created a "shared" experience for its viewers, an experience that results in *mythos*-fed fandom.

CATEGORIZING *COSMOS*

The rhetoric of science has been an important branch of rhetorical research and argumentation since the 1970s. Its popularity continues to grow as more and more scholars train their focus on the rhetorical approach and rhetorical significance of scientific texts. It's important to note that Aristotle did not believe that scientific communication was rhetorical because it relayed only truth; it was (as he saw it) not persuasive in nature. When rhetoric enjoyed a "rebirth" in the twentieth century, Robert K. Merton echoed this Aristotelian notion that scientific discourse is unquestionable, arguing that science is both universal (for everyone)[12] and communistic (shares the good it produces).[13] Merton did not see why science would need to persuade an audience because it produces only truth, is for the good of everyone, and is shared widely. As a result, he claimed, scientists do not need to manipulate their audience via persuasion.

However, in the 1970s, John Angus Campbell's "Charles Darwin and the Crisis of Ecology: A Rhetorical Perspective" argued that scientific writing indeed contains techniques of persuasion. Even though this seedling of "rhetoric of science" seems to have taken root from an investigation of a "public science" artifact, modern rhetoric of science scholars aren't in agreement whether popularizations of science are appropriately categorized alongside orthodox or "normal" science artifacts. Some rhetoric of science purists would exclude the investigation of an artifact like *Cosmos* from their taxonomical scheme within rhetoric. To some, *Cosmos* is an artifact that is more appropriately examined from a media analysis perspective rather than from a "scientific" perspective.

Even though there is disagreement within the field regarding whether the rhetoric of popular science should be classified within "rhetoric of science," several prominent scholars have indeed investigated the persuasive elements of scientific accommodations: Michael A. Overington, Leah Ceccarelli, Jeanne Fahnestock, S. Michael Halloran, Michael Leff, Alan Gross, Thomas Lessl, Greg Myers, Lawrence Prelli, and Carolyn Miller have all investigated popular science artifacts. However, most research in the rhetoric of popular science seems to focus on articles and not television. This subcategory of rhetorical investigation also doesn't seem to linger on the pop culture elements of the genre. As Marshall McLuhan famously noted, the medium of delivery in some ways *becomes* the message. Certainly, some assertions that have been made about the construction of *written* popular science texts may be applicable to an investigation of televised popular science. However, examining popular science *on television* will undoubtedly yield at least a few different insights.

It should be noted that analysis of the rhetoric of *Cosmos* has been done before but not at this length or to this level of concentration. Thomas M. Lessl published articles about the rhetoric of *Cosmos* in both 1985 and 1989. The earlier article, "Science and the Sacred *Cosmos*: The Ideological Rhetoric of Carl Sagan," argues that Sagan created a mythology out of science. Lessl notes that this mythology, combined with an audience of "average" people (nonscientists), makes this program difficult to categorize within rhetoric of science literature,[14] a stance that is no less valid now. Even though Lessl attempts some rhetorical explication of the program, he leaves his finding that *Cosmos* has a "mythical framework" only as an observation. He doesn't explore how it was built or muse on the potential impact on the program. Instead, he writes that *Cosmos* seems to transcend the educational program genre and that its myth-like construction is "reminiscent of numerous works of science fiction."[15] This intersection of science fiction, mythical frameworks, and popular science is the focus of this investigation into *Cosmos'* rhetorical structure.

Lessl isn't the only rhetorician ever to have written about *Cosmos* or to have used Sagan's rhetorical approach as an example. In one recent investigation of particular note, Lynda Walsh argues that mass media like television can help build *ethos* for science-focused programs, and she brushes on *Cosmos* and Sagan in illustration of her point. According to Walsh, a medium like television can combine with an already-existing predisposition of the public to trust science and its declarations,[16] making it a powerful medium for persuasion. This power, in addition to the requisite "encapsulation" of scientific subject matter on television,[17] means audiences likely aren't getting an entirely accurate view of the information, yet they are very likely to be persuaded by it. Walsh writes: "Technomorphic transformations . . . reduce science to symbolic parts of it."[18] These symbols can come to trigger the type of "inside jokes," or an in-the-know understanding, that Bormann describes as being *symbolic cues* in fantasy-theme analysis. In essence, Walsh connects this sense of being an insider with televised (popularized) science, making a natural bridge to fantasy-theme analysis of the genre.

Along with these ideas, it is important to note that most people's understanding of science comes from popularization. Few people outside the field read the scientific studies as published in scientific journals. The information they receive has, in some ways, been "predigested" or interpreted and then reflected out in a revised form. Underestimating the importance of public understanding of science or the role that these popularizations play in that understanding would be short-sighted. That said, however, the information reflected back is not *only* scientific. The magazines, newspapers, and television programs that transmit the revised scientific information have their own rhetorical motives, which do not replace the rhetorical motives of the scientists. Rather, their rhetorical motivations *layer on top* of that of the original authors.

For example, a scientist who has determined that a particular vitamin combination helps laboratory mice avoid obesity (despite overfeeding) writes a journal article with certain aims in mind. Those aims may include justifying funding for the project, using the success of the current project to levy funding for future research, and bolstering a performance or tenure file. Of course, the scientist also wants to persuade peers of the veracity of the findings.

If a news program picks up this information and runs a story, the language will be modified to fit a "slot" allocated within the program, thereby limiting the type and amount of information included in its report. While the news producers want their audience to believe the report, their primary concern is not with justifying funding, furthering the career of the scientist, or getting people to believe the findings. Their primary focus will be on getting people to watch. The motivation is about ratings; more specifically, it is about advertisers and advertising dollars. The news program may create a "teaser" that airs earlier in the day to entice audience to tune in again later. During the program, the story may also be teased as coming up after a commercial break. The main goal is to get the channel turned on and then keep the audience watching. Although this motivation is very different from that of the scientist, this new set of concerns does not erase or replace the concerns that were inherent in the original research. The original report likely used rhetorical techniques that, when examined, may reveal persuasive techniques that shape or color the original findings and how they were achieved. Changing the medium and presentation of the message does not erase the rhetorical nature of the original source.

Cosmos inhabits this intersection of scientific and commercial interest. While Sagan shared some of the motivations of the research scientist in the example above, he also needed *Cosmos* to attract an audience. Although the original *Cosmos* appeared on public television, it first needed to secure funding. The program's popularity allowed Sagan benefits and opportunities that he would otherwise not have enjoyed. For example, Sagan was the only member of "TTAPS"[19] (Richard Turco, O. Brian Toon, Thomas Ackerman, James Pollack, and Sagan) with enough freedom to speak openly about political issues like nuclear winter.[20] Part of this freedom was potentially linked to his job in academia (rather than at NASA or with defense contractors), but it was certainly linked with his visibility as a "public" scientist after *Cosmos*. The massive popularity of the program empowered him in a way that most scientists, especially researchers who work outside the public eye, never have the opportunity to secure. So, while *Cosmos* was not "purely" science, it also was not completely divorced from scientific concerns. It had rhetorical motivations that stemmed from media and entertainment—layered along with the rhetorical motivations of a scientist.

POPULARIZING SCIENCE

The context of a message's format (as well as its medium) shapes how society uses or applies that message.[21] In the case of popular science, and in this instance *televised* popular science, the context of its delivery centers on *entertainment*. That said, any program that appears on commercial television is, at the very least, passively selling products. Without the promise of selling air time to advertisers who are overtly selling their wares and services, the program will not be aired. David Tietge notes that the "corporate reality of the popular media" dictates that "ratings are everything."[22] Popularized science, whether in print or on the screen, is almost always motivated in some way by commercial concerns *because it is sold*. While some may argue that scientists also receive money for their work and so, theoretically, their work is also commercial, the fiduciary concerns are more massive, complex, and direct for popularizations than they are for any individual scientist.

Popular science has multiple, often highly diverse, audiences. In the same way that "normal" science articles need to persuade the editors of science journals to publish them,[23] so do popularizations have gatekeepers that may block distribution to their target audience. The avenues of distribution available for popularized science always involve some marketing component. Whether a state agency wants to distribute a health warning or a drug company wants to sell its latest miracle cure, popularized reflections of science need different types of financing and different forms of persuasion in order to be successful. While any message must first attract an audience before it can persuade an audience, the need to secure funding and sizeable audiences creates a need to cater to the sources of funding. In most cases of popularization, those "secondary" concerns are distinctly commercial. Commercial concerns may stem from entirely outside the rhetor's primary persuasive argument.

Television is primarily motivated by ad revenue, so Jeffrey Kirsch asserts that it must use "techniques that hold a viewer's attention sufficiently well to guarantee an audience for the advertisements."[24] Fiduciary concerns are at the core of this secondary audience: ethical or educational interests are not the deciding factors. The rhetorical message must be shaped to attract not only the audience, which the rhetor intends primarily to address, but also to attract and persuade an audience of financiers (gatekeepers) who need to fund the program's production, distribution, and/or publication. Unless a program is able to attract the necessary secondary audiences of funders and/or advertisers (and not just the primary audience of viewers), the program will most likely be scrapped.

Put another way, this secondary audience has controlling financial power and determines whether a message is created; that is, the secondary audience determines whether it even has a chance to *exist*. A program's ratings dictate the direction of corporate-driven television.[25] These ratings may focus on overall viewership, but they also measure certain demographics, like age. As Tietge asserts, without the "right kind" of ratings, "sponsorship quickly dries up, and from a rhetorical standpoint we must admit that without a significant number of viewers the message, no matter how important, will not be heard."[26] Without good cause to believe an advertising message will be delivered *as desired* to an audience of a particular size (and delivered, perhaps, to a *specific* demographic), financial support will likely not be secured.

These concerns about secondary audiences exist even if a program airs on public television, as *Cosmos* did, because PBS receives significant amounts of its funding from the Corporation for Public Broadcasting (CPB), corporate sponsors, and individual contributors. In 1980, 27 percent of the PBS's total funding (radio and television) came from the federal government, 37 percent from state and local governments, and 14 percent from subscribers and "auction/marathon."[27] PBS must be careful not to risk this state, local or federal funding by angering officials who hold the power to vote for or against funding. They also want to avoid controversy that may make individuals think twice about their personal "pledges" or contributions. The situation may be even more tenuous at for-profit cable channels or television networks, but their financial support comes from a variety of advertisers and it may, in some cases, be more easily (and quickly) replaced.

Intersecting Entertainment and Education: Fantasy-Theme Analysis

To attract a large audience, popularized science must include entertainment-focused elements. Although this book uses a neo-Aristotelian frame for rhetorical analysis, fantasy-theme analysis of *Cosmos'* rhetorical vision will provide some insight to the program's creation of *mythos*. The fantasy themes help to create the program's *mythos*, which will be explored in chapter five. This *mythos* creates a rhetorical vision, resulting in an initially appealing and ultimately long-remembered artifact.

Developed by Ernest Bormann in the 1970s, fantasy-theme analysis started out "as an humanistic method of rhetorical criticism"[28] and presented a hybrid of qualitative and quantitative analysis.[29] Bormann introduced it as a way of analyzing shared fantasies created through discourse, focusing initially on politics and religion. However, it also provides a critical lens for examining the connection between dramatization and social reality.[30] It must be noted that using the term "fantasy" does not call into question the *truth* of these beliefs. Instead, it focuses on a symbolic reality, or a reality that is often appreciated by and understood in the same way among a group of people. This distinction is particularly important when content is ostensibly scientific in nature.

The symbolic reality shared by a group can be triggered through a communal shorthand of some sort (which Bormann repeatedly equates with "inside jokes"[31]). The presence of this unified frame of belief demonstrates a common way of thinking among members of a "rhetorical community"[32]; it does not judge whether community adherents should or shouldn't believe. Instead, the term "fantasy" becomes a "technical term" in this context; it represents a way of categorizing "dramatizations that are nonfictitious as well as fictitious."[33] Symbolic reality is a rhetorical vision, or a "fantasy" of belief that has "chained out" and now pulls participants into psychological (or emotional) participation.[34] The participation does not need to be verbal or even purposefully expressed (although it can be): "facial expression, bodily posture, nonverbal evidence of rapt attention, and by laughter, moans, and groanings, sobbings [sic]."[35]

These triggers, or "symbolic cues,"[36] can take any form. A cue can be "a code word, phrase, slogan, or nonverbal sign or gesture. It may refer to a geographical or imaginary place or the name of a persona."[37] For example, a physical space can be a communal trigger. Memorials, especially when built upon ground sanctified through conflict or tragedy, can be very effective triggers. Even images of these memorials can be enough to activate shared memories or beliefs of people who share the same rhetorical vision. For one rhetorical community who shares the same symbolic reality, the Vietnam Memorial may trigger thoughts of young men drafted into service only to die far from home for a lost cause. In the same vein, mention of Hogwarts may trigger memories and shared symbolism for fans of J. K. Rowling's Harry Potter series.

Songs, movies, words, sounds, even smells can trigger these memories for members of rhetorical communities. However, those who stand outside the community are unlikely to fully understand the meaning embedded in the rhetorical trigger that can bring so many thoughts and emotions to the surface for others. Rhetors may intentionally try to create these triggers.[38] They may not be successful, however. Many never get the traction they need to capture the imagination or interest of a group large enough to create a shared symbolic reality.[39]

Fantasy themes are amoeba-like and unpredictable. Although rhetorical prowess can help to "spark" them, there's no guarantee anything will happen and there's no controlling where it will go once the symbolic reality ignites.[40] At the very least, Bormann notes that fantasy themes are perpetuated by "the mass media," who he refers to as "dream merchants."[41] In addition, Cragan and Shields emphasize the connection between current events and dramatizations.[42] They point out that fantasy themes of one group can conflict with the fantasy themes of another. They point to *Uncle Tom's Cabin* and *Gone with the Wind* as examples. They both contain fantasy themes regarding slavery, but these themes are in deep contrast to each other.[43] When a group forms a single frame of mind or way of thinking about or understanding something, then symbolic convergence has occurred.[44]

Rhetorical communities often form organically in response to the intersection of a *kairotic* moment in time and culture. As Hariman and Lucaites point out, images can be "suited to mass-mediated collective memory."[45] They make indelible impressions of a time; they carry meaning. In some ways, understanding and using the community's symbolic cues can help members identify one another. A slang term or phrase ("On Wednesdays, we wear pink") that may seem random or even nonsensical for some may open the memory of an entire experience and lexicon to others. As Bormann, Cragan, and Shields note, "Only those who have shared the fantasy theme to which the inside-joke refers will respond in an appropriate fashion."[46] Like an iceberg, the surface meaning of the message, the part that can be seen and understood by people outside of the community, is but part of the meaning that can be mined. Science fiction references trigger these elements of fantasy for the fans who have immersed themselves in the lore and have entered into the symbolic community. These symbolic cues reveal something about the "unfolding history" of the group; they often reflect the concerns, conflicts, and social mores of the time in which they are created.[47]

While Bormann briefly describes a process for fantasy theme analysis in his original article, this process focuses on beginning with the fantasy theme itself, then tracing the chain to find out where it originated and how the theme was manifested in various artifacts.[48] One flaw of the method, pointed out by some critics, is the difficulty in deciphering whether a rhetorical vision is created by an artifact or whether that artifact is only reflecting a set of beliefs or a community that already exists.[49] However, *even if* these rhetorical communities existed in some form previous to the introduction of the artifact in question, the communities will, at the very least, *change* because of new contributions to the fantasy symbolism that a new expression of the beliefs will bring with it. The "inside jokes," or symbolic cues, will naturally change because of the fresh reflection of the community's interests and/or values.

Fantasy themes are "filled with heroes, villains, emotions, and attitudes."[50] These heroes and villains guide the audience to project trust onto the narrator,[51] because the narrator is helping them to navigate through potentially "dangerous" waters. They can be identified by what Em Griffin describes as "increased energy within the group, by an upbeat tempo in the conversation, and especially through a common response to the imagery."[52] This sort of "increased energy" can be witnessed when a meme or video goes viral. When a quirk, like a dress that appears to be blue and black to some and gold and white to others, takes over the social media feeds of thousands of people, then a fantasy theme may be triggered or formed. Repetition of a fantasy theme provides evidence that a shared fantasy exists in a large group.[53] In the case of *Cosmos*, these include YouTube homages, references to Sagan's phrasing, and even reposted or satirized vignettes or images from the program.

For example, Darwin Deez's 2010 music video for his song "Constellations"[54] features the singer dressed in a Sagan-signature brown blazer with a turtleneck. While the song itself has little relevance to *Cosmos* (beyond mentioning stars and constellations), the Sagan reference is clear. Fewer than two weeks after "Constellations" was uploaded to YouTube, El Guincho released a video for the song "Bombay"[55] The introduction features the singer speaking lines from the first episode of *Cosmos*, "Shores of the Cosmic Ocean," in Spanish with English subtitles. The lines are some of the most famous from Sagan's introduction and include a reference to the iconic dandelion-seed Spaceship of the Imagination. While neither the rest of the video nor the song seems related to *Cosmos*, this introduction clearly references the thirty-year-old program. The coincidence of both videos being produced and released so close together, by two different artists thirty years after *Cosmos* first aired, illustrates the program's lasting effect on popular entertainment culture.

In order for something like this to have occurred, the program must have resonated with the creators of the videos—and these creators must have assumed some resonance was possible for their intended audience. This resonance was likely nostalgic, or based on emotion, which plays an important part in rhetorical visions. Once members of a group have psychologically connected to an artifact, they are then able to stoke their emotional connection to that artifact through something as simple as allusion,[56] as demonstrated by these videos. The chaining of symbolic reality-based fantasies links the audience together and pushes them to act in some way,[57] making the program inherently rhetorical.[58] Fantasy-produced social realties, which can be created and perpetuated by a variety of means (including electronic media), hold persuasive power.

Although the presence of emotional triggers may seem to tie fantasy-theme analysis to Freudian psychology, Bormann did not like—and even seemed to take offense to—this connection.[59] The audience's participation may not be intentional or necessarily voluntary, but Bormann compares their reactions to the way dreams may reveal information about the psychological state, or "agenda," of the group.[60] In other words, it isn't about the psychology of individual members of a group, but rather the shared psychology, goals, or values of a rhetorical *community*.

Investigating the symbolic cues, then, presents an opportunity for rhetorical analysis. How are audience members appealed to as individuals? How do these individual appeals then coalesce into group or shared fantasies? Exploring the neo-Aristotelian concepts of *ethos*, *logos*, *pathos*, audience, context, and purpose can assist in the search for answers to these questions. Bormann, Cragan, and Shields saw a natural link between neo-Aristotelian analysis and fantasy-theme analysis.[61] This book grounds itself in the connection between fantasy themes and the creation of *mythos*.

COSMOS: A PERSONAL VOYAGE

Cosmos has enjoyed decades-long popularity—popularity that could not have been reasonably foreseen when it was created. In order to trace the origins of audience engagement and the artifact's longevity, some chapters of this book will focus on neo-Aristotelian analysis using fantasy-theme analysis of the artifact using Bormann's key tenants of setting, character, and dramatic situation or action themes. Since *Cosmos* is a popular science artifact, this book will also use sources from the "rhetoric of science" to interrogate the program's rhetorical approach.

The presence of fantasy themes and the creation of *mythos* through intentionally created *kairotic* and *ethical* appeals makes the science presented in *Cosmos* more interesting to lay audiences. *Cosmos* is able to create a shared rhetorical vision by triggering pleasant associations with contemporary popular culture; by presenting a special effects-laden frame or forum through which the audience can view the journey and hear Sagan's arguments; by presenting a narrator who respects the audience, trusts them, who does not declare the audience's "ignorance" or talk down to them; and by demonstrating an allegiance with science fiction by deftly connecting thematically and visually to science fiction. *Cosmos* connects imagination with science. Without the power of the symbolic reality created through the program, the "lessons learned" in *Cosmos* would have eventually faded or been incorporated into common knowledge, their source of popularization forgotten. The symbolic reality of its fantasy themes creates added significance for the program, resulting in long-lasting popularity.

It's important to note before moving forward that this book will not examine whether the science of *Cosmos* is accurate. That's not to say that the science isn't correct; however, this consideration will simply not be part of this investigation.[62] While it is, of course, preferable if the science in a popularization is right, commercial concerns shift the focus away from accuracy and toward entertainment. Commercialism of science refocuses the priorities of the rhetors. The program and the science it contains have to be presented in a way that is interesting enough for people to watch. Chapter 5 will reflect briefly on how information may come to be presented in a way that is not precisely accurate. With that exception acknowledged, this book does not attempt to tackle every scientific fact in *Cosmos* that may potentially be inaccurate or skewed. This investigation is more interested with mechanisms of storytelling that may blur the lines between fact and fiction that create fantasy themes and a lasting popular *mythos*. It is not concerned with establishing absolute scientific accuracy or arguing whether either installment of *Cosmos* was the best translation of historical and scientific fact possible.

NOTES

1. David J. Tietge, *Rational Rhetoric: The Role of Science in Popular Discourse* (West Lafayette, IN: Parlor Press, 2008).
2. Adrian Malone, "Shores of the Cosmic Ocean," *Cosmos: A Personal Voyage* (Studio City, CA: Cosmos Studios, 1980), DVD.
3. Keay Davidson, *Carl Sagan: A Life* (New York: John Wiley, 1999), 318.
4. 1978's *Dragons of Eden*
5. Ibid., 262.
6. Ibid., 318.
7. Malone, "Shores of the Cosmic Ocean."
8. Thomas M. Lessl, "Science and the Sacred Cosmos: The Ideological Rhetoric of Carl Sagan," *Quarterly Journal of Speech* 71, no. 2 (1985): 177.
9. Lessl, "Science and the Sacred Cosmos: The Ideological Rhetoric of Carl Sagan."
10. Bill Carter, "'Civil War' Sets an Audience Record for PBS," *The New York Times*, September 1980, http://www.nytimes.com/1990/09/25/movies/civil-war-sets-an-audience-record-for-pbs.html.
11. Tietge, *Rational Rhetoric: The Role of Science in Popular Discourse*, 120.
12. Robert K. Merton, *The Sociology of Science: Theoretical and Empirical Investigations* (Chicago: University of Chicago Press, 1973), 270.
13. Ibid., 273.
14. Lessl, "Science and the Sacred Cosmos: The Ideological Rhetoric of Carl Sagan," 175.
15. Ibid.
16. Lynda Walsh, *Scientists as Prophets* (New York: Oxford University Press, 2013), 140.
17. Ibid., 145.
18. Ibid.
19. This famous group coined the term "nuclear winter" and helped the public to understand the destructive power of nuclear war.

20. Lawrence Badash, "Nuclear Winter: Scientists in the Political Arena," *Physics in Perspective* 3, no. 1 (2001): 88.

21. Barry Brummett, *Rhetoric in Popular Culture*, 4th ed. (Thousand Oaks, CA: SAGE, 2015), 213.

22. Tietge, *Rational Rhetoric: The Role of Science in Popular Discourse*, 112.

23. Philip C Wander, "The Rhetoric of Science," *Western Speech Communication* 40, no. 4 (1976): 227.

24. Jeffrey W. Kirsch, "On a Strategy for Using the Electronic Media to Improve the Public Understanding of Science and Technology," *Science, Technology, and Human Values* 4, no. 27 (1979): 52.

25. Tietge, *Rational Rhetoric: The Role of Science in Popular Discourse*, 112.

26. Ibid.

27. Alexander Field, "Table Dg172-180, Public Broadcasting Stations and Income, by Source: 1961–1997," *Historical Statistics of the United States Millennial Edition Online* (Cambridge University Press, 2015).

28. Ernest G. Bormann, "Fantasy and Rhetorical Vision: Ten Years Later," *Quarterly Journal of Speech* 68 (1982): 288–305, doi:0.1080/00335638209383614.

29. Ernest G. Bormann, John F. Cragan, and Donald C. Shields, "In Defense of Symbolic Convergence Theory: A Look at the Theory and Its Criticisms After Two Decades," *Communication Theory* 4, no. 4 (1994): 259–94; Bormann, "Fantasy and Rhetorical Vision: Ten Years Later." This method of analysis is part of what is now known as "symbolic convergence theory."

30. Ernest G. Bormann, "Fantasy and Rhetorical Vision: The Rhetorical Criticism of Social Reality," *Quarterly Journal of Speech* 58, no. 4 (1972): 400, doi:10.1080/00335637209383138.

31. Ernest G. Bormann et al., "Rhetorical Visions of Committed Voters: Fantasy Theme Analysis of a Large Sample Survey," *Critical Studies in Mass Communication* 1 (1984): 287–310, doi:10.1080/15295038409360038; Bormann, "Fantasy and Rhetorical Vision: The Rhetorical Criticism of Social Reality"; Bormann, "Fantasy and Rhetorical Vision: Ten Years Later"; Bormann, Cragan, and Shields, "In Defense of Symbolic Convergence Theory: A Look at the Theory and Its Criticisms after Two Decades."

32. Bormann, "Fantasy and Rhetorical Vision: The Rhetorical Criticism of Social Reality," 397.

33. Bormann, Cragan, and Shields, "In Defense of Symbolic Convergence Theory: A Look at the Theory and Its Criticisms after Two Decades," 262.

34. Bormann, "Fantasy and Rhetorical Vision: The Rhetorical Criticism of Social Reality," 397.

35. Bormann, Cragan, and Shields, "In Defense of Symbolic Convergence Theory: A Look at the Theory and Its Criticisms after Two Decades," 280.

36. Bormann et al., "Rhetorical Visions of Committed Voters: Fantasy Theme Analysis of a Large Sample Survey," 289.

37. Ibid.

38. Bormann, "Fantasy and Rhetorical Vision: Ten Years Later." Bormann sought to disprove Bales and show that fantasy themes could be formed with intention.

39. Em Griffin, *A First Look at Communication Theory*, 5th ed. (New York: McGraw-Hill, 2003), 9.

40. Ibid., 41.

41. Bormann, "Fantasy and Rhetorical Vision: The Rhetorical Criticism of Social Reality," 396.

42. John F. Cragan and Donald C. Shields, *Applied Communication Research* (Prospect Heights, IL: Waveland, 1981), 69.

43. Ibid.

44. Ernest G. Bormann, "Symbolic Convergence Theory: A Communication Formulation," *Journal of Communication* 35, no. 4 (December 1985): 128.

45. Robert Hariman and John Louis Lucaites, *No Caption Needed* (Chicago: University of Chicago Press, 2007), 2.

46. Bormann, Cragan, and Shields, "In Defense of Symbolic Convergence Theory: A Look at the Theory and Its Criticisms after Two Decades," 281.

47. Bormann, "Fantasy and Rhetorical Vision: Ten Years Later," 292.

48. Bormann, "Fantasy and Rhetorical Vision: The Rhetorical Criticism of Social Reality," 401.

49. Griffin, *A First Look at Communication Theory*, 43.

50. Bormann, "Fantasy and Rhetorical Vision: The Rhetorical Criticism of Social Reality," 398.

51. Sonja K. Foss, *Rhetorical Criticism* (Long Grove, IL: Waveland, 2009), 99.

52. Griffin, *A First Look at Communication Theory*, 38.

53. Bormann, "Fantasy and Rhetorical Vision: Ten Years Later," 294.

54. Darwin Deez, "Constellations" (YouTube, 2010).

55. El Guincho, "Bombay" (YouTube, 2010).

56. Bormann, "Fantasy and Rhetorical Vision: The Rhetorical Criticism of Social Reality," 398.

57. Bormann, "Fantasy and Rhetorical Vision: The Rhetorical Criticism of Social Reality."

58. Bormann, "Fantasy and Rhetorical Vision: Ten Years Later." Bormann notes, "My claims in 1972 were that fantasies are shared in all communication contexts, that there is a connection between rhetorical visions and community consciousness, that sharing fantasies is closely connected with motivation, and is an important means for people to create their social realities. These are no longer promising hunches. They have been supported by research results."

59. Ibid., 293. Bormann notes his aversion to Mohrmann and others making this connection in his footnotes.

60. Bormann, "Fantasy and Rhetorical Vision: The Rhetorical Criticism of Social Reality," 397.

61. Bormann, Cragan, and Shields, "In Defense of Symbolic Convergence Theory: A Look at the Theory and Its Criticisms after Two Decades," 268.

62. *Cosmos* was sometimes criticized for focusing too heavily on some issues and not enough on others. Some critics have also taken issue with the accuracy of the science presented. Other than a general examination of reflection versus representation, both of those concerns lay outside the parameters of this investigation.

Chapter 2

Cosmos' Context in Time

The opening sequence of *Cosmos* closes with a "flyover" from the top of the screen by Sagan's dandelion-shaped Spaceship of the Imagination. This flyover closely echoes the opening scene of the original *Star Wars* in which an imperial cruiser comes into view from overhead. This shot was groundbreaking and helped to cement that iconic movie as a work of master craftsmanship with awe-inspiring special effects. *Cosmos'* reference to the opening sequence of *Star Wars* not only makes a connection between the two for viewers but also establishes the differences between them. The imperial ship passing overhead in the movie adventure is threatening, especially given the way it dwarfs the rebel ship that it captures. The *Cosmos* ship is peaceful and poses no threat. It does not establish hierarchy but merely represents the potential of space exploration and therefore science.

These sorts of comparisons illustrate how Sagan and his team brilliantly used cultural touchstones to appeal to their original 1980 audience. When *Cosmos: A Personal Voyage* was produced, technology had advanced to a point where special effects could be mistaken for photographic reflections rather than artistic renderings. As a result, *Cosmos* was able to create a more realistic experience for its television viewers. Just as some popular entertainment touchstones of late 1970s/early 1980s culture (like *Star Wars*) have been held onto and revered by the following generations, so has *Cosmos* remained popular and come to be considered a classic. The shared fantasy the program created for viewers in their own living rooms had a lasting impact on many. The program introduced "a generation of young people" to science; some of them later went on to work for NASA.[1]

The timing of *Cosmos* critically affected its popularity. Had it been created a few years earlier, technology couldn't have supported the effort as effectively. Had it aired a few years later, competing channels and entertainment options may have made it less unique and captivating. In this way, *Cosmos* was able to take advantage of the rhetorical notion of "*kairos*" to increase its persuasive potential. Seizing a *kairotic* moment involves maximizing the impact of an argument by responding to its particular moment in time. An era's signature interests, fads, and obsessions can open a gateway if a rhetor addresses the contemporary pop culture. Television as a medium can be particularly effective in this regard.

If a text or artifact effectively responds to (or reflects) the concerns or interests of its contemporary audience, it then may even *become part* of the reflected culture. In a symbiotic circle of sorts, the artifact can simultaneously *speak to* and then *for* its era. *Cosmos*' enduring popularity through the years strongly suggests that it effectively responded to the *kairotic* code of its time and that it eventually became representative of it. Readily understood references to Sagan's vocal ticks ("billions and billions")[2] point to the program's interweaving—to some degree—with the popular culture fabric of its time. Although some organic interweaving certainly occurred, that does not discount the fact that the program—with intention—responded to current events and the sociopolitical climate of its era of production.

In the 1970s, frustrations about strained pocketbooks and economic inflation were nothing new. Spending public tax dollars on what may be seen as "luxury" or nonessential items would have been met with skepticism. Carl Sagan felt that it was especially important in this sort of economic climate to show the public how science affects the lives of everyone and the critical role science plays in improving society. People at large had to understand the contributions of science if they were to support scientific endeavors with their tax dollars. For Sagan, ordinary people had to understand the relevance of scientific exploration in their lives so they wouldn't reject it.[3]

At its core, *Cosmos*' mission was to move the public beyond education into *understanding* and support. With funding for sciences in jeopardy because of the strained economic climate, Sagan in particular wanted to persuade the public that government dollars spent on science benefit everyone, not just the elite and not just academics. *Cosmos* was constructed for resonance and persuasive impact, and those efforts were admirably successful.[4] Because the rhetorical concept of *kairos* shapes the foundational arguments in this book, this chapter reflects on the program's context in time. It will discuss the term *kairos*, then delve into 1970s popular culture, the socioeconomic climate of the time, and the ways these are reflected in Sagan's program.

THE RHETORICAL SITUATION AND *KAIROS*

To begin, it is important to look at the contextual "climate" of *Cosmos*' time to understand the ways in which the program appealed to its audience. As Bormann notes, an artifact must respond to the "unfolding history" that surrounds it: "the rhetoric must deal with anxieties aroused by times of trouble, by the evil defined within the social reality. The rhetoric must deal with changing circumstances and conflict."[5] *Cosmos* does that. It responds to and at the same time reflects the interests and concerns of its time. This chapter will provide the backdrop against which the "themes" of the program (setting, character, and action) play out. Without this anchoring in the popular culture and concerns of its time, *Cosmos* could not have hoped to resonate with its audience in the way it did. This resonance was powerful enough to make *Cosmos* a "classic" of popular science.

Before delving into the ways in which *Cosmos* responded to the context of its time, the rhetorical notion of *kairos* and what it means should first be explored. *Kairos* describes the convergence of multiple elements that come together in a moment of time, creating a situation that is primed for rhetorical intervention. Carolyn Miller applied the concept of *kairos* to science writing[6] and noted that responding to an "opening" requires acknowledgment and understanding of the events that converged to create it. *Kairos* is more than just "time"—it is also about *circumstance*. Time, people, places, interests, current events, method of message delivery—all combine to capture (or seize) a *kairotic* moment. It is the realization of perfect timing.

Richard E. Vatz frames the rhetorical situation as something that a rhetor responds to and thus creates. Rhetors shape situations by their emphasis or reaction. This process of discernment becomes a powerful means of persuasion.[7] For Lloyd F. Bitzer, however, either a changing of minds or inspiring of an action as *a result of the argument* is the factor that determines whether a situation is rhetorical. The audience's response, not the rhetor's attempts at persuasion, determines whether a situation is rhetorical.[8] So, even though the moment in time can be derived from occurrences and events outside of the rhetor's direct control, only the rhetor's success creates the rhetorical situation. Without rhetorical success—if no persuasion has taken place—then no rhetorical situation existed in the first place.

Like Vatz and Bitzer, James L. Kinneavy argues that the events surrounding a rhetorical discourse are vital in determining if an "opportune" moment exists for the rhetor to connect with his or her audience. Whereas Kinneavy defines *kairos* as focusing on "proper measure" and Plato defines *kairos* as being about the "propriety of time," Dale Sullivan argues that when these definitions are combined, they "suggest a third component, namely opportunity."[9] In short, the essence of *kairos* concerns identifying, creating, and responding to opportunity.

Popular Culture and *Kairos*

Creation of a shared rhetorical vision or symbolic reality relies on how audiences have "related to one another" and how they have "arranged themselves into social hierarchies."[10] Bormann's fantasy-theme analysis essentially examines the ways an artifact works to persuade its audiences to become believers[11] in the shared vision. In fact, Bormann actually uses the word "convert" as an adjective to describe the audience (converted vs. unconverted), lending a somewhat religious air to his discussion. Zealotry or fanaticism is at the root of any fan ("fanatic") culture. Fans value an artifact more than the average person; they are more "devout." Bormann also asks rhetors to think about the ways in which an artifact generates "a sense of community and cohesion from the insider."[12] These two subjects for examination—devotion and community—are closely linked. Community is formed through a shared sense of value and interest. It can be indicated through use of terminology or reference, like Bormann's "inside jokes" or symbolic cues.

The medium of *Cosmos*' delivery is crucial to the rhetorical content. Using "popular media as the vehicle" to present educational information changes how a program is constructed and how it reaches its audience.[13] The delivery shapes who is interested, who invests the time to watch, and what reactions these people have to *Cosmos*. Exploring how the use of television facilitates the persuasive impact of the message is a worthy avenue of investigation. It certainly opens the potential for creation of rhetorical vision that may not have been as successful in print alone. The popular entertainment of the 1970s indicates that audiences already enjoyed the symbolic reality created through the fantasy of science fiction entertainment. Audiences of 1980 were ready for the message Sagan delivered and for his method of delivery. The rhetorical situation presented itself and *Cosmos* seized it.

SHARED FANTASY AND FANDOM

Shared fantasy is the essence of fandom. *Close Encounters of the Third Kind* can provide an example of this type of *kairotic* transcendence and the creation of a rhetorical vision. The movie was released in 1977 and was one of the most financially successful of its decade.[14] The special effects created a realism that could only be dreamed of in the black and white UFO "B-movies" of the 1950s. Color made those mid-century movies slightly more "high tech," but the effects were often hokey and unbelievable. *Close Encounters* seized its *kairotic* moment by responding to cultural interest in space and the paranormal. Yuri Geller, zodiac signs, and other parascience fads were considered popular during this decade; UFO abduction stories were one iteration of these types of interests. *Close Encounters* was able to tap into those interests while at the same time becoming part of the popular culture of the day. The effects and the slow unfurling of the storyline lent immediacy to the contents of the film. At the same time, the filmmakers didn't simply update the same type of story for a new time—they created a lasting pop culture *mythos* about alien "visitors." The film is closely associated with the 1970s cinema as one of the first modern blockbusters. Today, *Close Encounters of the Third Kind* is considered a classic.

In some ways, being considered a "classic" is an acknowledgment of how successfully an artifact tapped into its *kairotic* code and created a rhetorical vision. Remakes of these artifacts are rare.[15] Occasionally, additional installments appear years later, but it can be very challenging for these texts to trigger the nostalgia of the original while still tapping into the *kairotic* code of their own time. For example, *Terminator 2: Judgment Day* was in released 1991, seven years after *The Terminator*. The sequel stayed true to the original's storyline, but it updated the technology used to tell that version of the story. Like the first *Terminator*, *T2* was ultimately very successful. With an estimated budget of $102 million, it grossed nearly $200 million.[16] *The Terminator* was made with an estimated budget of $6.4 million and made $38 million.[17] According to IMDb, both enjoyed successful opening weekends: $4 million for *The Terminator* and $31 million for *T2*. The *Terminator mythos* resulted in the common catchphrase, "I'll be back"; *Terminator 2* resulted in the catchphrase, "*Hasta la vista, baby*." The popularity and easy recognizability of these sayings demonstrate the types of symbolic cues that result from shared appreciation for an artifact. Repetition by the audience or other artifacts indicates a shared interest in the symbolic reality created by it.

In contrast, *The Godfather* and *The Godfather: Part II* were both very successful and appeared only two years apart. *The Godfather: Part III* was released sixteen years after second part. This third installment was much less successful than the first two. Each movie's net profit went down, from $129 million for the original to only $12 million for third part.[18] While the reasons for this difference may be debated, the interim sixteen years certainly resulted in a different cultural interest and attitude than received the original installments. The original entered the contemporary lexicon with phrases like "make him an offer he can't refuse"; "it's not personal . . . it's strictly business"; and "sleeps with the fishes." The third installment only had one ("Just when I thought I was out, they pull me back in")—and it appeared in the trailer, as did a good deal of footage from the earlier films. Once again, presence of these easily recognized and often-used sayings indicate participation in the shared rhetorical vision. However, the rhetorical vision attempted by the third installment was highly dependent on the symbolic reality created by the first two movies. Given the intervening expanse of time, it could be argued that the situational context had slipped by and *kairotic* transcendence was not possible for the final installment.

Cosmos also had an intervening expanse of time between installments. *Cosmos* 2014 was created and aired more than three decades after the original. Signature phrases associated with the franchise (like "billions and billions"[19]) still trigger memories of the original program and its host. After thirty years of advances in technology and science, developing a new program would seem logical and, frankly, simpler. Doing so would allow greater creative flexibility; there would be no need to "follow in footsteps." The program could innovate and create on its own. However, the popularity of the original carries rhetorical power. Connecting back to the juggernaut original *Cosmos* increased the potential for the new program to attract an audience since it could then carry with it the recognizable *Cosmos* pedigree. Popular science can benefit from association with a recognizable brand just as reboots, remakes, and sequels (or prequels) of fictional programs can benefit from their popular predecessors of the same name. Unlike the *Godfather* franchise, each installment of *Cosmos* was ultimately successful[20] despite a wide gap in time and in the concerns of its audience. To capture its *kairotic* moment, the original *Cosmos* responded to and even embraced contemporary concerns of the 1970s and early 1980.

ADDRESSING THE SOCIOECONOMIC CLIMATE

Cosmos delivered sizable audiences because it addressed its rhetorical context so well. It may be hard to imagine in the politically and ideologically polarized America of 2017 that a program, which challenged life-guiding constructs such as religion (like *Cosmos* did), would reach such a large audience or have such a long-lasting impact on it. To understand how, we must explore the socioeconomic reality of the 1970s, including popular opinions about defense spending and policies; scientists and science; the economy; the environment; science education; and entertainment trends.

Weapons, War, Science, and Scientists

The space program of the 1960s enjoyed a good deal of positive public support.[21] Unfortunately, science in the 1970s was closely associated with war, and for good reason: through the end of the Cold War, weapons and defense drove and advanced scientific research, whereas after the cold war, capitalism did.[22] At the time *Cosmos* was produced, Daniel Kevles reports that "National defense accounted for about 50 percent of the total [$2.3 billion Research and Development budget in physics for 1977]; the space program, especially the space shuttle, for another 15 percent."[23] The contrasting amounts demonstrate the spending priorities of the US government at the time. Unfortunately, the discrepancies also cast a negative light on scientists themselves.

As Kevles notes in *The Physicists: The History of a Scientific Community in Modern America*, the race to the moon helped to energize the public toward a common "scientific" goal in the 1960s. However, once that goal was met, public interest in science and space exploration hit a decline.[24] During this time, public opinion of science and scientists was also influenced by works such as Lewis Mumford's *Pentagon of Power*, which claimed that after 1940, a "scientific-military-industrial machine" had emerged.[25]

In general, the public perceived the hard sciences as being run by favoritism. According to Kevles, it didn't help public perception of science or scientists that then-President Lyndon Johnson's home state of Texas seemed to reap disproportionate rewards in science program funding. At one point, 40 percent of research proposal reviewers came from only ten universities: "Scientific research, the acrid wisecrack went, was the only pork barrel for which the pigs determined who got the pork."[26] The perception of science as the recipient of government money—not as a contributor to society—was damaging. Being equated with politics, political parties, and favoritism painted science not as the "hero" of the space program but as an arm of the Establishment.

Science was increasingly being seen by a financially struggling public as a powerful and corrupt entity, which gobbled up resources and was not to be trusted. NASA's refusal to share a report with congress "following the flash fire that killed three astronauts," claiming that the report "might destroy the 'intimate and confidential' nature of NASA relationships with its contractors"[27] only served to compound negative public opinion. Preserving a relationship within that "scientific-military-industrial machine"[28] (rather than satisfying the public's desire to learn what happened) further demonstrated that science's concerns seemingly did not focus on society or the good of all. Naturally, this refusal added to already-negative public perceptions regarding "scientific elitism" and its confluence with "industry and government" in relation to Vietnam.[29]

The American public had begun to view science as being in partnership with big business and not as a tool or "helper" of the general public at large. Defense contracts were rich and the American people saw the war in Vietnam as a cash cow for science: in turn, they viewed science as increasingly motivated by money and fiduciary opportunity. In short, the "social needs of the nation"[30] were not being met but the financial needs of the war machine were. Public opinion held that a game of political favoritism was being played with tax revenue. These funds, in the view of some citizens, could have gone to support social programs that benefited a large number of people but instead was benefitting only a few. Through "an arms race which employed nearly half the world's scientists,"[31] the public had become jaded to science as a means of warfare development and a tool of government agenda-building.

As a case-in-point, a 1977 article entitled "The King Kong of Federal Waste" openly mocks some of the research projects the government endeavored to fund,[32] deeming them "unworthy." The main thrust of the article is that investing tax dollars in studies on Frisbees, tricycles, perspiration, acorns, and toads seems ill-advised during a time in which the government was borrowing money just to remain solvent.[33] At about the same time, *The Sciences* ran an article entitled "The Visible Scientists" focusing on the "uneasy relationship" between science and the public. It blames this discomfort on science's "antiquated concepts" about public disclosure.[34] It also argues that since society is suffering from "technological ills," science must update its methods of communication.[35] In other words, the article is requesting more transparency from science. Popularization would be one way to accomplish what is being requested.

Even the promotional coverage for *Cosmos* notes the contemporary negative view of science. Frederick Golden, in the 1980 *Time* cover story about *Cosmos*, writes that science's public image was widely acknowledged to be poor: "A decade ago . . . [science] was unpopular, even in disrepute . . . [It] was being blamed for much that was wrong with the world."[36] While public opinion of science had improved somewhat by the time *Cosmos* premiered in September 1980, the improvement was focused in the area of medical research.[37] Even though applied physics had given society "digital watches and pocket electronic calculators [and] television photos transmitted from spacecraft on Mars," public opinion of physics and research in applied physics did not seem to improve.[38] During his promotion of the program, Sagan touted that the "primary objective of *Cosmos* was to show that science is a delight and to end people's artificial alienation from it."[39] Ending this alienation would be key in repairing science's poor public image.

The Economy

The early 1970s was particularly problematic for the US economy. Longtime Congressional Representative Henry Reuss noted in a speech, "1973, I'm afraid, must go down in history as pretty close to being a disaster."[40] To support his claims, he notes that the cost of living had increased during that year at a rate of about 9 percent—but pay increases averaged only 5.5 percent, meaning that "average" Americans (who were most likely to suffer from these economic fluxuations the most) were worse off financially than they had been the year before.[41] His speech takes aim at poor policy decisions, dishonest political dealings, and the energy crisis, but not at science.[42] Unfortunately, he projects that 1974 would continue to see high inflation rates of about 8 percent or 9 percent.[43] Reuss argues that half of the US people were "alienated because they feel they're being ripped off by a society in which big government and big business looks out after its own interests and leaves to the average person . . . the short end of the stick."[44] The discontent of the American people with the economy is clear.

Reuss wasn't alone in bemoaning the particularly difficult economic time. About that same time, *The Saturday Evening Post* declared that "everyone" anticipated that "1974 [would] be a banner year for dollar dilation."[45] The same article also notes that no end to inflation was in sight. Government intervention and attempts to lessen negative impact (like price controls) hadn't helped the situation. In fact, oil prices quadrupled in a single year because of a false "shortage" created by OPEC (Organization of Petroleum Exporting Countries), which had conspired to manipulate and control its production.[46] These issues were at their worst in the early and mid-1970s, but at the beginning of 1980, the tough times were not yet over.

According to the March 17, 1980, issue of *Time*, the United States was suffering under an enormous 18 percent per year inflation rate. This article, descriptively entitled "The Economy: 'Scary' Inflation Rages on, and Everybody is Getting Angry," also reports that declaring bankruptcy as a nation seemed plausible to some.[47] Many Americans wanted the government to cut back on funding expenditures. Taken in perspective, the millions spent on space exploration seemed to be a bad idea in the context of looming national fiscal failure. In illustration, the article notes that US Treasury Note interest rates had increased from 10 to 15 percent since September of the previous year (1979); unemployment was around 6 percent but wholesale prices rose in February 1980 by "a compound annual rate of 19.69 percent"; and interest on business loans was up to 18 percent.[48] US citizens, on the other hand, wanted cuts—but not necessarily on programs from which they benefited directly—a circumstance that made politicians reluctant to cut social programs.[49] These sorts of preferences and demands were bound to set science up as an "easy-to-cut" outsider since the benefits it produced weren't necessarily immediate and could be harder to see.

Cosmos was the result of Sagan's desire to change public perception of science so the public would no longer see it as an area easily cut. One article describes this quest to change public perception of science as "desperate," noting that scientists "need to get Americans interested and excited about supporting future science programs."[50] One of the scientific casualties of the bad economy of the previous decade was Halley's Comet; *Physics Today* lamented: "A rendezvous mission to Comet Halley . . . was not funded."[51] It would seem natural that the United States would not let such a highly visible opportunity pass by, given the name-recognition and history of the comet. However, the public's concern about funding NASA missions such as this one reflected a difficult economy; investment in domestic affairs that would help struggling families seemed like a better investment.

The Environment

The late 1970s was a time of contention regarding governmental spending on scientific endeavors and a time of financial struggle for many American citizens, but it was also a time of concern about the environment. Organizations were attempting to change the perspective that humans are on the top of the evolutionary ladder, placed there by a divine being who would always ensure human dominance. According to Lincoln Cushing, the first Earth Day, which was in 1970, "signaled a dramatic escalation of public perception about environmentalism and ecology."[52] It expressed a growing concern over pollution. In fact, "the late 1960s saw the first public efforts to ban the widely-used [sic] pesticide DDT."[53] William R. Brown and Richard E. Crable note that the public at large tended to place culpability for environmental woes on "corporate America for a polluted society."[54] As a result, these companies launched "educational movements" geared toward convincing the public that industry was not, in fact, at fault for pollution and that it was doing its part to control the problem: "in the 1970 issues of Time, Businessweek, and Newsweek alone, there appeared a total of 289 pages of environmental advertising."[55] Exposure to the topic of pollution and emphasis on environmental awareness was not just in news stories; it was even in the advertisements.

At the same time, rhetoricians were also turning their eyes toward environmental issues. John Angus Campbell, arguably the first rhetorician to examine the rhetoric of science, wrote in 1974, "In our current crisis of the environment we are looking for perspective on man and nature."[56] Public concern was raised regarding pollution and the environment in the early 1970s: "Many young people and even noted writers variously claim that either western science or western religion or both advance a view of man's relationship to the environment that is harmful and spiritually bankrupt."[57] The environmental movement had, for some, stirred a desire not just to repair or stop damage to the planet but also to find the sources of it—whether those sources were corporations or something else—perhaps in an effort to more effectively stop and reverse the damage.

This situation created great optimism for attracting an audience for science-based programming. Judy Klein reported that television programmers believed that "the alarm exhibited in response to the events at the Three Mile Island nuclear plant [indicated] widespread and growing concern about the effects of modern technology."[58] The Three Mile Island accident, which occurred in 1979, was not the only environmental disaster that was causing concern; according to Bruce Lewenstein, "several major news stories of the [previous] decade were 'science stories' [that focused on] Mount St. Helens [1980], Love Canal [mid-1970s], and the 'superfund' for environmental clean-up [1980]."[59] Daniel Kevles reveals that even as far back as 1969, "pollsters discovered that Americans ranked the space race far lower in importance as a national problem than water and air pollution or job training and poverty."[60] Public concern and priorities were focused on immediate economic and environmental concerns.

Science Education

With the Public Broadcasting Act of 1967, Lyndon Johnson paved the way for establishing PBS. The original intention was to provide programming that was not motivated by ratings and that would serve needs that weren't otherwise being met.[61] In other words, diversity was a foundational goal of public television in its inception. The PBS board was originally made up of lay people who wanted to increase the quality of programs available on TV. However, by 1975, Lawrence Wenner notes that public TV was already being controlled by the elite, not the "average citizen."[62] He argues that average people would have had similar levels of experience as other people on the board and fewer extraneous issues to vie for their attention.[63] Instead of focusing on creating an advisory board made up of the key demographics for public television, Wenner implies that the board didn't really think of (nor even consider) involving the public. Rather, it sought "eminent" people like "corporate execs and former government appointees" over educators, artists, or people who worked in civic affairs.[64] Although the government acknowledged a need for noncommercial, educational television, the name "public television" was chosen when "educational television" seemed off-putting to potential audiences.[65]

Advanced education at this time was somewhat rare. In the early twentieth century "very few students continued their education beyond the elementary level,"[66] although the situation by *Cosmos*' time was not quite as bad. Unfortunately, Jacob Bronowski "failed" in his goal in getting the public to care about and understand science[67] with his 1973 program, *The Ascent of Man*. Even though science education reform had been attempted repeatedly since the 1950s, no initiative worked as well as had been hoped.[68] American education was lacking in the key area of science and no solution had yet been found.

The separation of the public from a full grasp of the importance of science and its impact on their lives was well underway by the late 1970s and early 1980s. Reports showed that there was "a significant decline of scientific ability among American students," and at the same time, budgets were suffering from "'a fifteen-year shrinking of the U.S. commitment to excellence in science, math, and engineering'."[69] As a result, even adults did not have a level of science literacy that allowed them to understand the complexities of science or space exploration. In 1982, Allan Bromley recognized that this lack of understanding was very serious: "Over 80 percent of our citizens receive their last exposure, if any, to mathematics and science during their high school years."[70]

Although the general public may have lacked a full understanding of science, the public still valued knowledge about science and tended to be very responsive to stories about science.[71] If television as a medium and science as a field "translated" science for public consumption—that is, if both television and science nurtured the development of scientific accommodations—Harold Davis argues that both could then have a great "impact on public understanding."[72] What was still missing was perhaps broad public *interest* in science, which could not be whetted when the general perception of science was not one of adventure or advancement but instead one of crookedness and the Establishment.

Entertainment

Popular culture can arguably serve as the barometer of a society's interests and values during a snapshot in time. Popular entertainment in the 1970s did not shy away from addressing big political questions and controversial topics. According to Kirsten Lentz,[73] television shows of that decade were making a concerted effort to reflect typical "life experiences" and challenges of the time. Shows like *Maude*, *Good Times*, and *All in the Family* were exploring "taboo" topics like sex, race, and politics. The lead character in *Maude* debated having an abortion. Edith Bunker, the lead female character in *All in the Family*, escaped an attempted sexual assault. In *Good Times*, the Evans family struggled with crime and racism as they raised their family in what they called "the projects."[74]

At that time, the considered "quality" of a show depended in part on how—and if—the program reflected the grittier aspects of reality that had largely been ignored in 1960s bubblegum programming like *The Brady Bunch*.[75] Norman Lear, MTM Enterprises, and others "[transformed] the situation comedy, making it more complex and more responsive to the social and political changes resulting from the civil rights and black power movements and the burgeoning feminist movement."[76] Television pushed the boundaries of programming to reflect a life that was closer to that of some viewers, not idealized or sugar-coated as it had been in the past. As Lentz points out, the producers "expected young audiences to respond positively to television programs that adopted the logic of and issues associated to relevance."[77] As programming executives emphasized "relevance" in their programming, it became "tolerated" among viewers, and it could be said that addressing contemporary political and social issues in entertainment media even became expected.

This trend toward social argument in entertainment may have been connected to the very real social and economic concerns that permeated the lives of the contemporary audience: "historians who write about 'quality' and 'relevance' television tend to understand the two as part of the same general discursive and economic shift in the 1970s."[78] Television programs also embraced what Jason Mittel calls "narrative complexity," which he defines as a practice in which programs emphasize plot over character relationships or drama. Mittel notes that "this move toward complexity dates to the late 1970s and early 1980s."[79]

In addition, the public had experienced a transformational moment in film when *Star Wars* was released in 1977. A character-driven movie filled with action and emotion, a movie about war, adventure and salvation, *Star Wars* echoed some of the "everyday" internal conflicts and desires for adventure that audiences were experiencing in their own lives. For example, the "Force" was so popular that some fans encouraged George Lucas to create a religion based on it.[80] The massive pop culture-permeating audience reception of *Star Wars* indicated a hunger for adventure and fantasy, not just the life-reflection or contemplation inherent in relevance programming on TV.

Star Wars was set in another galaxy in which space travel was a common proposition and it also spoke to the *kairotic* code of the 1970s American audience. It provided the comfort of classic storytelling within a science-fantasy setting. Adding to the fantasy was its opening line, which indicated that all events documented in the film occurred "a long time ago." The implication, of course, is that the movie documented another galaxy's *history*, filled with spaceships and interplanetary travels. It was a space movie that wasn't futuristic. *Star Wars* presented adventure wrapped in fantasy, but it provided fodder for reflection about potential societies in other galaxies—societies much more advanced than our own. Those advanced societies would look at the spaceships we saw on the screen and view them in the same way the world and time of the 1970s looked at the horse and buggy. *Star Wars* didn't simply bring audiences a daydream: it presented to them an archetypical "good versus evil" story presented in unique and intriguing ways.

By the 1970s, television had been commonly available long enough to have been researched for audience reception and reaction. These studies helped to guide network executives and program producers in making production decisions, not just for entertainment-centered shows but also for popular science programs. To that end, Jeffrey Kirsch, the director of the KPBS Science Center in 1979, wrote an article entitled "On a Strategy for Using the Electronic Media to Improve the Public Understanding of Science and Technology." In it, Kirsch asserts that knowledge about target audiences could be used to create popular science programs that capture large audiences. He defines "the dilemma of science communication" as a need to package content using "terms and symbols" that are able to elicit "resonant responses in the audience."[81] This "packaging" should be done "along formulaic lines" in the form of characters "based on the stereotypes and shared fantasies of the American public."[82] Kirsch's intent is to tap into the current interests and knowledge of the American viewing public to create science programming that interests them. The rhetorical approach to popular science on PBS at this time is thus not something that is later determined *only* through analysis of the rhetorical artifact—but these remarks make clear that PBS's rhetorical approach in regard to popular science was formed *with intention.*

Science had been difficult to package and present in the early years of television, but by 1980, new technology and improvements in television production allowed science-centered news segments to live up to their potential.[83] Out of this new-found production value came one of commercial television's first "successful" primetime science-focused programming efforts in the form of *Walter Cronkite's Universe*, which executives hoped would "look like *Star Wars* and sound like 'Sixty Minutes' [containing] segments on breaking science news, investigative pieces and theoretical pieces."[84] This science-news focused program aired first in the summer of 1980, just before *Cosmos* premiered in September. However, it is interesting to note that *Cosmos* was already into production and was being discussed in promotional articles when *Walter Cronkite's Universe* was still an unnamed concept searching for the seed money needed to produce a pilot.[85] It's possible that Cronkite's program was produced as a preemptive response to *Cosmos'* potential popularity. Regardless, contemporary producers of popular science programs likely realized connecting science with entertainment would improve their chances of success.

Walter Cronkite's Universe, like *Cosmos*, had its special effects associated in the press with *Star Wars*. *Time* described *Cosmos* in a markedly similar way: "It features special effects rivaling those in *Star Wars*: computer animation, scale models and painted backdrops as dazzling as anything ever attempted on television."[86] Other promotional literature for *Cosmos* pushed the comparisons to science fiction even further by touting,

> Following the lead of many recent science fiction films, 'Cosmos' [sic] uses special effects lavishly. There are more than 70 special effects sequences using such cinematic devices as model animation, band and front projection, chromakey, explosions, tank shootings, mattes and traveling mattes, and new technology effects that use motion control systems for multiple-layer compositing, according to PBS.[87]

PBS also described *Cosmos* as "the first TV program to blend state-of-the-art special effects with nontechnical language in an effort to take some of the scare out of science."[88]

Connecting to the most popular movie franchise in history (up to that point) was key to promoting *Cosmos*. Viewers were likely to have seen *Star Wars*, perhaps multiple times, so finding that they may be able to have a similar experience in their own homes (in the mostly pre-VCR/DVD/pay-per-view era) would likely have been attractive to them. Marketing materials focused on promising audiences another such space adventure. Sagan wanted to bring science to the public in a more interesting, engaging way using "innovative wizardry."[89] Jeff Rovin reported in September 1980, "Sagan is convinced *COSMOS* [sic] will enlighten a large audience because the program is entertaining rather than pedantic."[90] Others in the media noted that the special effects were sure to be "an impressive lure to film buffs."[91]

In addition, a Screen Actor's Guild strike that occurred in the fall of 1980 delayed the premieres of many scripted programs until weeks into the new television season. Finally, on September 25, network execs and SAG reached an agreement that would allow for the "long-delayed new fall TV schedule [to] be in place by the last week of October or the first week of November."[92] While it is impossible to know for sure the effect this strike had on *Cosmos*' ratings, it is worth contemplating whether this strike also contributed to creating a *kairotic* moment for the program, which premiered on September 28 of that year. Had the regular TV season begun as scheduled, would some of *Cosmos*' audience been syphoned away? In an age when available channels were limited, perhaps *Cosmos* was able to capitalize on the lack of other popular programming. Since *Cosmos* was the most popular program on PBS up to that time, these circumstances may have combined with the adroit way in which Sagan addressed contemporary concerns and entertainment preferences of the time.

PBS Audience Demographics in 1980

Who was the primary audience for *Cosmos*? Ratings data for PBS programs of that era is scarce. While they've held a Nielsen contract since the 1980s, those contracts covered only "12 sample weeks" out of a year.[93] Not until 2009 did they expand their contract to fifty-two weeks a year. It's unclear which episode(s) of *Cosmos* were rated (or how) during its initial run, but it remained the highest rated PBS program until *The Civil War* aired ten years later.[94]

In lieu of comprehensive program-specific information, this examination will use "typical" PBS demographic data from 1980. In a published PBS status report, S. Young Lee discloses that by March 1980, about 90 percent of US households had the potential to receive public TV.[95] During this same window, "the size of the public television audience was at an all-time high, having increased consistently since March 1975."[96] PBS audiences were also watching PBS in prime time; those numbers increased by about 9 percent between March of 1979 and March of 1980.[97]

In terms of demographics for the typical public television audience, Young reports that about 58 percent of the PBS audience made an income of $15,000 [approximately $43,500, adjusted for inflation] or more; in contrast, 52 percent of the total TV viewing audience was making the same amount.[98] TV audiences for PBS were most concentrated in the Northeast (28 percent) and the South (24 percent), with the West Central (19 percent), Pacific (17 percent), and East Central (13 percent) regions following behind.[99] Young also notes that PBS households were more likely to have children—4 percent more likely—and their children were more likely to be younger than that of the "average" TV-watching household.[100] Percentages of white (89 percent vs. 88 percent) versus nonwhite households (11 percent vs. 12 percent) were not remarkably different between PBS and general TV-watching households.[101]

The most significant differences in public TV audiences as opposed to total TV audiences seems to be in education. Households that watched public TV at this time were usually "headed by someone with one or more years of college, while more than one-quarter (27 percent) of the household heads had less than a high school education."[102] Only 27 percent of PBS viewers were likely to have not completed high school, whereas 33 percent of the total television audience didn't finish.[103] Audiences who had finished high school were about equal (35 percent for PBS and 34 percent for total TV viewership), according to Young; however, PBS's audience was about 6 percent more likely to have attended "one or more years of college."[104]

Cosmos addressed the interests and concerns of its time, but that message was received by approximately the same numbers of nonwhite households as the average TV audience. The largest segment of its audience likely lived on the east coast, made a slightly higher-than-average income, had young children, and enjoyed the benefits of at least a high school education and possibly some college. Audiences for commercial television would have been likely to live in the Northeast and South and had a high school education, but they would have been less likely to have children (particularly young children) and to have attended any college. They would have been more likely to have dropped out before finishing high school and to come from the East Central region of the United States. In the end, PBS viewers had larger incomes and were at least slightly more educated on the whole. Since public television was created to address diverse needs and to fill "gaps," it seems that this divergence between the national averages and PBS averages indicates that, at least to some extent, a different audience *was* being catered to. However, the diversity wasn't great as the overall picture of television audiences reflected in Young's report is mostly consistent across both groups.

COSMOS IN CONTEXT

To understand how *Cosmos* created a *kairotic* opening for itself, considering the ways in which the program responded to its largest audience, their interests and imaginations, is important. To begin, the Spaceship of the Imagination takes an unusual shape: as a ripe dandelion seed, it does not seem to conform to any preconceived expectations about how a spaceship should (or would) look. Yet, it reflects an understanding of the time and responds to its contemporary rhetorical context. In a time concerned with weapons build-up and war, Sagan's ship reflects the transportation not of death but of life. Rather than being equipped for defense and attack, Sagan's ship intends only discovery. In some ways, it is like the "olive branch" from the biblical story of Noah and the ark. This approach, from the perspective of peace, sets an "unthreatening" tone for the program from its opening credits.

This peaceful vision of the future was not exclusive to *Cosmos*. The common fantasy of human evolution often includes this type of tranquility. The ship's interior is reminiscent of the initial cool calm of HAL from *2001: A Space Odyssey*. The mission in that film is presented as critical, worthy of every sacrifice. Sagan's Spaceship of the Imagination is not homicidal like HAL, nor prone to psychological breakdown; it is the means of human evolutionary advancement. Just as Dave, the "Star Child," is depicted as the next step in human evolution at the end of *2001*, so the Spaceship of the Imagination shows its audience the wonders of universal evolution, the birth, and death of stars. Similarly, the dandelion seed presents an unexpected portal of locomotion, like TARDIS in *Dr. Who*. It is not a typical vehicle, and that uniqueness helped it to become memorable and iconic. In other words, its representation assisted in creating visual reference for the program—a symbolic cue.

The dandelion spaceship referenced public frustration over defense spending and the prevailing perceptions of science and scientists in several ways. For example, the ship is presented as a dandelion seed, a natural "vehicle" of life. Sagan holds a seed in the first episode and lets it go in the wind. Afterward, where it travels can no longer be controlled by him. In this way, the seed represents both freedom and adventure as well as "release." It can be associated with exploration, learning, comfort, intelligence, and peaceful adventure. Since the ship has no weapons, it is also entirely peaceful. It looks toward the future creation of life; it does not harken back to the Old West via side arms (as did *Star Trek*). On the interior of the ship, Sagan sometimes sits and sometimes stands. At all times he is engaged in the adventure with us, whether it is "calibrating" instruments in some way or marveling at the images appearing outside the ship. He is an active participant, sharing with the audience an adventure that has cost them nothing but their time. The white, pointed interior is reminiscent of a Christian church or cathedral—a place of charity and communion but also of consistency. With an economy that was changing so rapidly, having a relatively unchanging place of respite would have been a relief to many. Referencing such a place through the ship design may have elicited some of that same feeling from the audience.

While a dandelion seed may have seemed an odd choice for a spaceship, it demonstrated an allegiance with audience members who may have been struggling in a difficult economy. Dandelions are free and proliferate widely—and they are edible. During hard times, dandelion greens can be cooked or preserved, the blossoms fermented into wine, and the greens, blossoms, *or* roots steeped in hot water for tea. Some even say that dandelions can convey certain health benefits. They are a flexible and valuable resource for those who need one. They are hearty survivors, difficult to extinguish even under trying conditions. Inside the ship, the single chair for Sagan, our "captain," demonstrates leanness. Some comfort, but not needless comfort, has been provided. Expenses haven't been bloated for unnecessary luxuries. Even Sagan's clothing speaks to consideration of finances. The textured tie and jacket are not overly "posh"; they are reasonably stylish but not over-the-top or too sleek. He is dressed like a contemporary late 1970s/early 1980s man, not like an astronaut, lab scientist, celebrity, or alien. His dress is not futuristic. It fits within a common,[105] contemporary image of an ordinary person.

The audience first sees the dandelion seed on the Monterey, California, coast. This seashore becomes a point for departure, but it is also metaphoric. Just as that coast represents the "edge of the cosmic ocean,"[106] so does the dandelion represent the shared vessel for exploration. It travels under the power of imagination—not fossil fuels. The ship represents a way to travel to the next step in human evolution as a species, as the audience explores the vastness of space. This seed is part of nature, a natural construct that cannot be "manufactured." The ship's lack of destructive power and its association with a life-spreading seed make it a fitting environmental ambassador to the stars in 1980.

According to Dale Sullivan, entertainment in education is "achieved by the artful mixing of the familiar with the novel."[107] The dandelion is one of the most familiar symbols *Cosmos* could have selected. It is extremely common on earth and the seeds spread on the wind. By portraying a seed as the ship, Sagan is giving familiarity to the unfamiliar, comfort along with the potentially uncomfortable. These elements work together to provide comfort enough that audiences might relax. If they relax, they may trust the program to take them to new places. Even as it tries to teach them things that may challenge their beliefs, they will have the constancy of a comforting environment to return to. It will be familiar but not boring.

The Spaceship of the Imagination is the original "Magic School Bus,"[108] taking the audience on a field trip to the universe. The ship can then enjoy the benefit of all the good memories viewers may have related to the fun of climbing aboard the school bus (to explore new places and to learn new things). These memories are likely related to fun and escape. Being the product of imagination, the ship distances itself from association with any coldness, hardness, or rigidness that might otherwise be associated with science. Imagination is not controllable by others, so the impression of self-efficacy within the journey may be present. At the very least, if the ship is made of "our" imagination, the audience and Sagan are creating the vehicle together. The audience is contributing to the journey, rather than being a passive passenger.

Sagan sought public support of scientific research. By tapping into the *kairotic* code of the time, he sought to evoke the emotional responses that would make his audience more receptive to his arguments. Referencing popular culture and socioeconomic concerns of an artifact's contemporary audience ends up doing more than simply attracting that audience. In the case of *Cosmos,* it ends up creating another rhetorical "pull" or angle for persuasion.

Cosmos responded to the economic, educational, and social concerns of its audience. It intrigued viewers by offering new information but the program packaged it within comfortable imagery and symbolism. It mixed the "familiar with the novel"[109] to create a shared fantasy and rhetorical symbolism that could unite a rhetorical community of fans. As chapter 3 describes, the *ethos* established by Sagan's host and guide further reinforces these references to contemporary concerns and interests.

NOTES

1. Keay Davidson, *Carl Sagan: A Life* (New York: John Wiley, 1999), 331.
2. Sagan never said this phrase in *Cosmos*, although it is closely associated with him and the show. It was actually said by Johnny Carson in satirizing his frequent guest.
3. Adrian Malone, "Who Speaks for the Earth," Cosmos: A Personal Voyage (Studio City, CA: Cosmos Studios, 1980).
4. Thirty-four years after the original program first aired, a rebooted version of *Cosmos*, called *Cosmos: A Spacetime Odyssey*, appeared on primetime commercial television.
5. Ernest G. Bormann, "Fantasy and Rhetorical Vision: Ten Years Later," *Quarterly Journal of Speech* 68 (1982): 292, doi:0.1080/00335638209383614.

6. Carolyn Miller, "Kairos in the Rhetoric of Science," in *A Rhetoric of Doing: Essays on Written Discourse in Honor of James L. Kinneavy*, ed. Stephen P Witte, Neil Nakadate, and Roger D. Cherry (Carbondale: Southern Illinois University Press, 1992), 310–27.

7. Richard E. Vatz, "The Myth of the Rhetorical Situation," *Philosophy and Rhetoric* 16, no. 3 (1973): 157.

8. Lloyd F. Bitzer, "The Rhetorical Situation," *Philosophy and Rhetoric* 1, no. 1 (1968): 3.

9. Dale L. Sullivan, "Kairos and the Rhetoric of Belief," *Quarterly Journal of Speech* 78 (1992): 318.

10. Bormann, "Fantasy and Rhetorical Vision: Ten Years Later."

11. Ibid., 402.

12. Ibid.

13. David J. Tietge, *Rational Rhetoric: The Role of Science in Popular Discourse* (West Lafayette, IN: Parlor Press, 2008), 131.

14. "Close Encounters of the Third Kind," *The Internet Movie Database*, n.d., http://www.imdb.com/title/tt0075860/.

15. When remakes or updates are made, especially if much time has passed between the original and the remake, they infrequently receive the same overwhelmingly positive reception of the original (no matter how popular).

16. "Terminator 2: Judgement Day," *The Internet Movie Database*, n.d., http://www.imdb.com/title/tt0103064/.

17. "The Terminator," *The Internet Movie Database*, n.d., http://www.imdb.com/title/tt0088247/.

18. "The Godfather," *The Internet Movie Database*, n.d., http://www.imdb.com/title/tt0068646/; "The Godfather: Part II," *The Internet Movie Database*, n.d., http://www.imdb.com/title/tt0071562/?ref_=nv_sr_2; "The Godfather: Part III," *The Internet Movie Database*, n.d.

19. This phrase was not actually said in *Cosmos*, but rather it originated from *The Tonight Show with Johnny Carson*. Carson coined the phrase in parody of his sometime-guest.

20. *Cosmos* 2014 did not enjoy the same level of success as the original, however.

21. Fred Jerome, "On Science and American Society," *Environment* 23, no. 7 (1981): 20.

22. Tietge, *Rational Rhetoric: The Role of Science in Popular Discourse*, 21.

23. Daniel J. Kevles, *The Physicists: The History of a Scientific Community in Modern America* (New York: Alfred A. Knopf, 1978), 424.

24. Ibid., 399.

25. Ibid., 400.

26. Ibid., 397.

27. Ibid., 401.

28. Ibid., 399.

29. Ibid., 401.

30. Ibid., 404.

31. Malone, "Who Speaks for the Earth." This quote is from Ann Druyan's introduction to the year-2000 re-issue of the program on DVD.

32. Nick Thimmesch, "The King Kong of Federal Waste," *The Saturday Evening Post*, January 1977, 66+-66+.
33. Ibid.
34. Rae Goodell, "The Visible Scientists," *The Sciences* 17, no. 1 (1977): 6–9, doi:10.1002/j.2326-1951.1977.tb01494.x.
35. Ibid.
36. Frederick Golden, "The Cosmic Explorer," *Time*, October 1980, 65.
37. Kevles, *The Physicists: The History of a Scientific Community in Modern America*, 423.
38. Ibid.
39. Kristin McMurran, "His Cosmos a Huge Success, Carl Sagan Turns Back to Science and Saturn's Rings," *People*, December 1980, http://www.people.com/people/archive/article/0,,20078104,00.html.
40. Henry S. Reuss, "State of the Economy: Unemployment and Inflation," *Vital Speeches of the Day* 40, no. 10 (1974): 295.
41. Ibid.
42. Ibid.
43. Ibid., 296.
44. Ibid., 297.
45. "Inflation Here to Stay?," *The Saturday Evening Post*, April 1974, 46.
46. Gerald A Pollack, "The Economic Consequences of the Energy Crisis," *Foreign Affairs* 52, no. 3 (1974): 452.
47. "The Economy: 'Scary' Inflation Rages On, and Everybody Is Getting Angry," *Time* 115, no. 11 (March 1980): 16.
48. Ibid.
49. Ibid.
50. Marilynn Preston, "Carl Sagan's 'Cosmos' Will Mean Worlds to You," *The Chicago Tribune*, September 1980, A16.
51. "President Requests More Support for R&D at NSF and NASA," *Physics Today* 31, no. 3 (1978): 104.
52. Lincoln Cushing, "Posters about Ecology and the Environment Before and During the 1970s," *Electronic Green Journal* 1, no. 30 (2010): 1.
53. Ibid., 3.
54. William R. Brown and Richard E. Crable, "Industry, Mass Magazines, and the Ecology Issue," *The Quarterly Journal of Speech* 59, no. 3 (1973): 260.
55. Ibid., 260–61.
56. John Angus Campbell, "Charles Darwin and the Crisis of Ecology: A Rhetorical Perspective," *Quarterly Journal of Speech* 60, no. 4 (1974): 442.
57. Ibid., 443.
58. Judy Klein, "The Medium Gets a Message," *Science News* 115, no. 22 (1979): 365.
59. Bruce V. Lewenstein, "Was There Really a Popular Science 'Boom'?," *Science, Technology, and Human Values* 12, no. 2 (1987): 31.
60. Kevles, *The Physicists: The History of a Scientific Community in Modern America*, 398.
61. Dollie Cole, "Public Television," *The Saturday Evening Post*, 1974, 60–61, 83.

62. Lawrence Wenner, "Citizen Involvement in Public Television," November 1975, 13, http://files.eric.ed.gov/fulltext/ED123676.pdf.

63. Ibid.

64. Ibid., 14–15.

65. Cole, "Public Television."

66. Morris H. Shamos, *The Myth of Scientific Literacy* (New Brunswick, NJ: Rutgers University Press, 1995), 41.

67. Ibid., 103.

68. Ibid., 132.

69. Quoted in Marcel Chotkowski La Follette, "Multivariate Perceptions: Reflections on the 147th Annual Meeting of the American Association for the Advancement of Science," *Science* 6, no. 35 (1981): 35.

70. D. Allan Bromley, "The Other Frontiers of Science," *Science* 215, no. 4536 (1982): 1037.

71. Harold L. Davis, "Needed: More for Public Understanding," *Physics Today* 28, no. 6 (June 1975): 88, doi:10.1063/1.3069041.

72. Ibid.

73. Kirsten Marthe Lentz, "Quality versus Relevance: Feminism, Race, and the Politics of the Sign in 1970s Television," *Camera Obscura* 15, no. 1 (2000): 44–93.

74. The programs present their living arrangements as a racial and economic "ghetto."

75. Ibid.

76. Ibid., 46.

77. Ibid., 60.

78. Ibid., 47.

79. Jason Mittell, "Narrative Complexity in Contemporary American Television," *Velvet Light Trap* 58 (2006): 32.

80. Lucas declined the opportunity to begin a religion based on "the Force."

81. Jeffrey W. Kirsch, "On a Strategy for Using the Electronic Media to Improve the Public Understanding of Science and Technology," *Science, Technology, and Human Values* 4, no. 27 (1979): 54.

82. Ibid., 53.

83. Klein, "The Medium Gets a Message," 61.

84. Ibid., 365.

85. "Two Science Series Developed for TV," *Science News* 114, no. 16 (1978): 263.

86. Golden, "The Cosmic Explorer," 63.

87. "TV Test Tube: PBS Brews Two Science Series," *BioScience* 30, no. 10 (1980): 265.

88. Jeff Rovin, "Carl Sagan's Cosmos," *OMNI*, September 1980, 105.

89. Ibid., 106.

90. Ibid.

91. Sam Maronie, "Science to the People," *Starlog*, October 1980, 24.

92. John Carmody, "Tentative Agreement in Actors' Strike," *The Washington Post*, September 1980, https://www.washingtonpost.com/archive/lifestyle/1980/09/26/tentative-agreement-in-actors-strike/a92eb96b-1848-4d26-af1f-ee5e906a57e8/.

93. Kevin, "PBS Gets Rated," *Adweek*, December 2009, http://www.adweek.com/tvnewser/pbs-gets-rated/27101.

94. Bill Carter, "'Civil War' Sets an Audience Record for PBS," *The New York Times*, September 1980, http://www.nytimes.com/1990/09/25/movies/civil-war-sets-an-audience-record-for-pbs.html.

95. S. Young Lee, "Status Report of Public Broadcasting, 1980" (Washington, D.C., March 1981), http://files.eric.ed.gov/fulltext/ED205205.pdf.

96. Ibid.

97. Ibid.

98. Ibid., 38.

99. Ibid.

100. Ibid.

101. Ibid.

102. Ibid., 37.

103. Ibid., 38.

104. Ibid.

105. His clothing leans somewhat to the professorial, but it is far removed from the stereotypical "scientist in a white laboratory jacket."

106. Adrian Malone, "Shores of the Cosmic Ocean," *Cosmos: A Personal Voyage* (Studio City, CA: Cosmos Studios, 1980).

107. Dale Sullivan, "Exclusionary Epideictic: NOVA's Narrative of Excommunication of Fleishmann and Pons," *Science, Technology, & Human Values* 19, no. 3 (1994): 290, http://www.jstor.org/stable/690026.

108. *The Magic School Bus* is an educational program from the 1990s. School children and their teacher are transported through space on a yellow school bus to learn about science.

109. Ibid.

Chapter 3

Ethos and the Popular Science Persona

To attract and retain the large audiences needed to justify the millions of dollars spent on producing *Cosmos*, Carl Sagan had to compose this presentation of popular science strategically. As discussed in chapter 2, "*Cosmos'* Context in Time," when *Cosmos* was produced, scientists were seen by the public as tools for the Establishment. Many people viewed them as belonging to a profession in which the "pigs" were in charge of the "pork."[1] According to George Gerbner, fictional representations on television depicted scientists as "relatively less attractive, fair, sociable, warm, tall, young, or peaceful, but very smart [although in some instances] less rational and stable and much more violent than other characters."[2] Scientists' public image centered on only one good trait (intelligence) but on many lesser traits that were generally not desirable. Sagan needed to have his "host" character display admirable, likeable characteristics in order for the audience to accept him and his message. Projection of elitism or superiority would likely have resulted in audience rejection of the program.

As a result, Sagan created a *dramatis personae*,[3] or "character," that contributes toward the construction of the program's symbolic reality. He presents his character in *Cosmos* as a guide or navigator, an explorer, and a teacher/advisor—not just for the program, but for the human race. He becomes a benevolent, all-powerful being that can be relied upon for truth. He moves the story forward, taking the audience anywhere in space or time. He gains trust by exhibiting vulnerability. He is not a master but a mentor. Sagan's persona or character allows him to become the "pied piper" who entices the audience to follow along on a cosmic journey, even when it ventures into unknown or controversial territory.

Lynda Walsh notes that dramatic visual representations (like that of *Cosmos*) leave audiences with the impression that scientific discoveries are "dramatic acts of Divination."[4] She notes that the interior of Sagan's Spaceship of the Imagination is "church-like," a fitting environment for a scientist/priest who is "exhorting his flock to wonder."[5] Sagan becomes an "oracle" to his viewers because of the "awe" or wonder they feel as they witness what is revealed on their television sets.[6] The pull of persuasion is great when it springs from the Divine. Audiences come to believe they are really seeing what is going on, even though what they are seeing is manufactured or altered for entertainment purposes. The "window" through which audiences are let into the inner workings of science is, according to Walsh, "fundamental to the prophetic *ethos* in the media age."[7]

Cosmos' first episode, "Shores of the Cosmic Ocean," demonstrates the prevailing character themes surrounding Sagan's *Cosmos* host. Sagan's divinity is reinforced by the critical entrenchment of his message with the "vehicle" of television. In the end, the journey overseer who guides the program's cosmic exploration is a carefully crafted creation, not necessarily fully reflective of Carl Sagan's own personality. This reality is as undeniable as it is predictable. Only for educational or other "truth-based" programming would such an assertion (that the "personality" projected on-screen is not same as that projected off-screen) be made.

Without Sagan's character, the rhetorical vision of the program is unlikely to have been realized. This character "provides ultimate legitimatization of the drama,"[8] as defined by Bormann. Sagan's character is able to take advantage of his status as an oracle, as a person of divinity, omnipotent, omniscient, and omnipresent—like a god. His presence helps the *Cosmos* journey resonate with audiences. They are able to rely on Sagan for safety and for knowledge. He becomes their divine[9] host and they become the disciples of his rhetorical vision.

FORMATION OF *ETHOS*

Whereas an issue of "credibility" might appear to be tied to the reliability or trustworthiness of the speaker, Aristotle describes *ethos* as a *projection* of character,[10] not a *reflection* of character. Narrators or "hosts" of popular science television programs must be appealing or personable. If they aren't, audiences will not tune in or continue to watch. In many (if not most) instances, this projection is not reflective of the true personality or bearing of the host. The identity presented can be defined as a "persona," as described by Paul Newell Campbell.[11] Although *persona* is different from *ethos*, the two work in tandem, creating a character that is appealing or interesting and perceived to be trustworthy or credible.

Because the audience must perceive a speaker favorably before they will open themselves to any argument,[12] character in popular science is important. By agreeing to hear an argument, the audience is thereby conferring power to the speaker. In this way, *ethos* represents an exchange of power. The audience must trust the speaker's knowledge and sincerity before considering or even hearing the speaker's arguments. A speaker cannot "create" or "take" *ethos*—it must be given willingly by the audience. Speakers must reflect appreciation for the audience's popular culture to earn *ethos* and thereby gain this power.[13] To show that the speaker shares the values of the culture, he or she must address the audience in a way that takes into account the historical context. As explored in chapter 2, the rhetor must "speak" to the code, the language of the time.

In addition to the educational background or knowledge level of the audience, the speaker must take into account the sociopolitical climate that produces and reflects interests, attitudes, and concerns in the audience. Once the rhetor has gained trust by demonstrating or acknowledging an allegiance to the values of the audience, the audience will open themselves to hear the speaker's message. The audience must first decide that they can believe and trust the rhetor before "tuning in" or opening themselves to persuasion.

According to Baumlin and Baumlin, *ethos* isn't simply an acknowledgment of shared experience and values; *ethos* requires the audience to "project" their own internalized images of "authority and trustworthiness onto the speaker."[14] Referencing the important touchstones of an audience makes them more willing to receive a speaker and therefore the argument being presented. If the speaker appears trustworthy, knowledgeable, and interesting enough to earn the audience's attention, then the audience may respond positively to the speaker, opening themselves to the persuasion that will inevitably be embedded in the message the speaker will deliver.

PERSONA IN POPULARIZATIONS

Popularizations do not follow the same standards of communication established by the scientific community for academic works because the rhetorical situations are very different. According to Robert K. Merton, physicians and lawyers have a clientele that may shape what those fields produce, but a scientist has no such limitations.[15] For Merton, science is "pure" and explores for knowledge's own sake. The truth behind that assertion may be debated, but *popular* science *does* have a clientele, customers that it has to please. As a result, content must be presented in a way that both attracts a broad, sizable audience and inspires financiers to support its production. Popular science has to adapt its messages both to an "intended" audience and to an audience of financial backers.

As a result, those who control the program funding must be persuaded that the topic and its presentation will produce a certain number of viewers, for example, before the message can even be created. In short, popularizations have to "move" the audience in different ways than does orthodox science. Programs must seek to "resonate" with the values and experiences of the audience[16]—and these values and experiences are bound to lie outside the boundaries of science. Therefore, no static system for vetting or presenting popularized science is available, nor would one be effective.

Popularized science can't be relayed according to a set pattern. Information about the interests and concerns of the contemporary audience force popular science as a genre to form appeals and characteristics that speak to the audience of the time. In addition, an element of drama is arguably necessary to attract and keep an audience that is less likely to spend time with a "school lesson" than they are to watch an entertaining, interesting program from which they also can learn some science.

To help in the endeavor of adding drama and interest to science as well as to show a connection with the concerns and interests of the contemporary audience, most televised popular science enlists a narrator or host to guide the audience through the program. While this person may be an actual scientist and benefit from some of the inherent *ethos* resulting from belonging to that collective, the speaker also must be entertaining or interesting. Therefore, the formation of a "character" is necessary. This formation of character follows the Aristotelian notion that projection and not reflection is what is necessary in establishing *ethos*. The persona exhibited on-screen by the host of a popular science program no more reflects that person's "true" personality than dramatic actors' characterizations are reflective of theirs.

More than simply providing a guiding hand to assist viewers along their journey, a host "character" may also embody a character theme that contributes to the program's overall rhetorical vision. This host can become a symbolic cue that members of the program's rhetorical community can use as a touchstone for identifying each other and for invoking memories of the show. Sagan's character in *Cosmos* is still often satirized, demonstrating the longevity and popularity of the program. The idea of Sagan, the representation of his benevolent deity, still resonates with fans.

Whereas the creation of persona or character is vital in popular science presentations, it is a rarity in traditional science. Walter Weimer views the "persona of ideal science" as cold, without specific, identifiable interests, and consumed by logic.[17] In popular science, cold, emotionless detachment will lose an audience very quickly. Personality and personal appeal are critically important. While Aristotelian rhetoric acknowledges that persona is a projection and not a truthful reflection of the speaker's character, in science this dissembling contrasts with the discipline's need to be accurate, honest, and descriptively exhaustive.

Even so, persona in scientific works can also be very effective in securing an audience. According to Michael Halloran, Watson and Crick's groundbreaking and often-republished announcement about the structure of DNA was written in a "consciously contrived style."[18] Halloran notes that they present themselves as "genteel" but "in the flesh, they were obstreperous and irreverent."[19] Watson and Crick create a persona not just for other scientists (who also needed to accept their work), but also for the journalists who would read their announcement and would write stories about it. Watson and Crick's fabricated rhetorical persona helps them to extend the reach of their science beyond the limits of their field, according to Halloran, to take ownership of it. The creation of such an expertly crafted persona, one that was purposefully meant to appeal to a nonspecialist audience while also communicating to other scientists, was groundbreaking. In a way, Watson and Crick created a fantasy-based character theme that was anchored in their personas as authors and scientists. By portraying themselves as humble "good guys," rather than the competitive and manipulative people they were, they became heroes within the story they presented.

Watson and Crick's projection of persona influenced the press's reception of their announcement, and as a result, it may have rewarded them with more public notoriety than they otherwise would have had. This notoriety for Watson and Crick effectively eclipsed the contributions of Wilkins, Stokes, and Wilson as well as of Franklin and Gosling, the scientists who wrote the two papers that accompanied the original publication of "A Structure for Deoxyribose Nucleic Acid." According to Halloran, these other two articles were not written using the same persona and so were left out of some subsequent reprintings.[20] It could be that the lack of appealing persona may have made them seem much more dull. In promoting science, character-based fantasy themes can be a useful rhetorical tool.

SAGAN: REALITY VERSUS FANTASY

In the first episode of *Cosmos*, "Shores of the Cosmic Ocean," Sagan presents himself as an educated "everyman" with the ability to share wondrous information with the audience. However, according to biographer Keay Davidson, Sagan's own personality was reportedly abrasive, insensitive, and belittling,[21] as well as "brash" and "aggressive."[22] During the filming of *Cosmos*, "Sagan seemed to lose his sense of humor and became increasingly brusque, even petulant at times."[23] He clashed with director Adrian Malone and he was dealing both with a contentious divorce and the failing health of his father. On his shoulders he carried concern for his mother, who would be lost without his father. Halfway into production, the "stresses on the production [of *Cosmos*] . . . were driving it close to collapse."[24] Nowhere in Sagan's depiction of his amiable scientist guide is any of this stress apparent to average viewers. His tone of voice and patterns of speech remain consistently gentle, patient, and kind with his signature pronunciations and emphases.

To his credit, Sagan was always a talented and generous teacher. Davidson reports that Sagan "almost always displayed patience and good humor when addressing lay people."[25] While his colleagues may have found the representation of Sagan in *Cosmos* difficult to watch (perhaps because it conflicted with the personality they knew), the public responded very well. The show became what Davidson described as a "cultural triumph" with "textual clarity and visual dazzle. It was to television science documentaries what *2001: A Space Odyssey* had been to science-fiction (sic) cinema: a new standard to shoot for."[26] Some critics, of course, were skeptical and interpreted a good deal of ego in the presentation of the host's persona.

Sagan's expressions and signature style of speech were "off-putting" to some.[27] Some contemporary critics also begrudged Sagan's spotlight in the program. As the *only* consistent face and voice throughout the 13 one-hour episodes, Sagan's dominant presence was at times perceived as ego. One contemporary critic wrote, "His image is never ending . . . . In Sagan We Trust."[28] Other critics also grumbled about his seeming lack of "humility."[29] Some wanted less emphasis on emotional appeals and would have preferred the "more detached urbanity of an Alistair Cooke."[30] Even with these criticisms, though, critics noted that his presence—not *a host's* presence, but *Sagan's* presence in particular—was vital to the program. One critic noted, "I don't know if the series could have been done without him, but it's hard to believe we needed this much of him."[31]

While his "persona troubles some viewers,"[32] the success of *Cosmos* would have been unlikely (if not impossible) if the majority of viewers did not find the persona of the program's host to be pleasant, appealing, and trustworthy. According to a *Time* magazine cover story, Sagan "without a doubt [was able to make] science as palatable as the apple pie."[33] The program is so closely associated with Sagan that it is logical to assume "the phenomenal commercial success of the series . . . was due largely to Sagan and his quirky yet compelling persona."[34]

As a credentialed scientist, Sagan was able to take advantage of the collective *ethos* of science, but he also had to avoid being grouped with orthodox scientists as one of the "elite." An attempt to walk a thin line between "everyman" and "scientist" may have contributed to some critics' impression of him as "sanctimonious"[35] while others responded more positively. Prior to *Cosmos*, Sagan had made several late-night TV appearances with Johnny Carson that gave him some cultural cachet as someone worth watching. Still, Sagan could not rely on these initial impressions to sustain an audience through the entire thirteen-episode program. He needed the audience to connect with him as the program's host so they would connect with the subject matter. The information needed to be delivered in an interesting way to depart from school lessons and make the program an adventure. Sagan accomplished this feat through the environment in which he first introduced "himself," or, his persona, as someone who is powerful yet unthreatening, knowledgeable, and trusting.

Sagan's character contributes to the symbolic reality created by *Cosmos* in several ways. First, he satisfies an innate human yearning for God. Sagan represents not just a guide, but a higher power. His association with the Divine provides an impression of privileged access. This divine being is wise and loving, lulling the audience into comfortable participation alongside Sagan. These traits are reminiscent of *Star Wars*' Obi-Wan Kenobi, an older man who guides the movie's hero through space adventures to his destiny. Kenobi stands to gain nothing from helping Luke and even sacrifices his life to serve an example for this young man who barely knows him. He teaches Luke a new way of existence, a new sort of "religion" that permeates and controls all life—not life on a single planet, but life throughout the galaxy.

Sagan becomes the *Cosmos* audience's Obi-Wan in revealing the great wonder of the universe and the secrets that it holds. He serves as a protector and savior. Of course, his savior persona reflects his projected divinity. By the end of the thirteen-episode program, he has warned the audience of "sins" committed in the past (i.e., the destruction of the Library of Alexandria) and has warned of the fragility and importance of life. By the final episode, he shows the audience a sudden "silencing" of Earth. Only by taking heed of Sagan's advice can humanity be saved from doom. The illustration of the Drake equation, which charts out the probability of life on other planets, also contributes to this sense. The lesson of the equation is simple: if only humanity can live past its dangerous predilection for war, it can advance to greater things beyond this planet. Sagan's advice is instructional but meant as a life lesson. It's designed to encourage viewers to take control of their own evolution and destiny. This is the program's central rhetorical vision: to save all life on earth and possibly in the universe.

Second, Sagan's character satisfies a typically American thirst for adventure. He demonstrates this power from the first episode.[36] He takes viewers from their comfortable, familiar Earth to the "edge of the known universe." Then he brings them slowly back home, narrating the universal wonders that come across the screen as they go. This journey is reminiscent of science fiction movies that were popular at the time—and Sagan is the one who can facilitate the audience's participation in their own adventure. Even though Sagan took the program far away, he still kept the audience safe. Since viewers are able to remain in their living rooms, they can enjoy the comfort of their surroundings. Anything "foreign" that enters into the room on their television sets is controlled by Sagan. His calming narration explains what they see and they are never left to be surprised or to flounder. The audience is allowed to wonder and to enjoy the journey—from the safety and familiarity of their own homes.

Last, the character allows Sagan to present vignettes in easily digestible form. If Sagan the scientist were to teach the lessons, they would undoubtedly be more complex and technical. As presented, they are short, narrated for clarity, and often provide re-enactments, examples, or illustrations. The tales that his character spins are cautionary tales as well as tales of fortitude or failure. These stories are grounded in history, in some ways reminiscent of the biblical stories his viewers may know. Told by this character, the cosmos' tale is an epic journey about creation, life, strife, and turmoil—and it also offers peace and salvation.

These stories and their delivery further cement Sagan's place as a godlike savior. In one episode,[37] Sagan reveals his own hometown as the Bensonhurst neighborhood of Brooklyn, New York, and he takes viewers to one of his childhood classrooms. The students are excited and happy, seeming to revel in Sagan's presence. Sagan also shares a story about a life-changing trip to the library there when he was a little boy. He asked the librarian for a book about "stars" and received one back on celebrities. After he clarified, the librarian brought him a book on astronomy. These vignettes are presented as the humble roots of an ordinary person. However, they also connect him even more strongly to a Christ-like figure: an exalted being with humble human beginnings but possessing a divine purpose and godly power.

"The Shores of the Cosmic Ocean"

The opening ten minutes of *Cosmos*' first episode establishes Sagan's persona for the audience. It illustrates the fantasy theme his character is intended to portray, his vast but unthreatening power, his intellectual prowess and his trustworthiness. Although the entire episode is geared toward setting the stage for the twelve episodes that follow, these key initial minutes establish the audience's willingness to confer *ethos* to their guide, largely because of his presentation of persona. This section will first describe these minutes in detail and then will break them down to describe how they work rhetorically. It should be noted that Hulu, the website from which text selections were originally transcribed from captions, aired the 2000 reissue of *Cosmos*, which includes a two-minute introduction by Ann Druyan. Druyan, a *Cosmos* cowriter, indicates the importance and longevity of *Cosmos*, reaffirms the science, and compliments the program as "prophetic." This introduction would, of course, not have appeared during the original airing in 1980, but it is notable to mention because all minute markers in these scene descriptions include these additional two minutes.

After the Druyan introduction, as the first episode of *Cosmos* begins, reddish-hued stars and galaxies of different shapes are presented on a black background. Then "Cosmos" appears on the screen, which is soon replaced with "by Carl Sagan." Afterward, appearing alone on the screen are the words, "A Personal Voyage." Soon thereafter, a spaceship in the form of a radiant dandelion seed flies over head. Then, the name of the episode is presented: "The Shores of the Cosmic Ocean."[38]

This opening montage (which is repeated for each episode) then transforms to an image of waves curving, crashing to shore. We hear Sagan's voice (speaking in a calming tone) before we notice him in a long shot at 3:04, walking, barely perceptible, toward the cliff. He is walking from the left of the screen toward the center, and as he turns toward the camera, it zooms in slowly while maintaining a still-significant distance. He is clearly speaking, his hands moving with his words, but he has no recognizable facial features or facial expressions. At 3:12, he speaks using dreamy, poetic language:

> The cosmos is all that is or ever was or ever will be. Our contemplations of the cosmos stir us. There's tingling in the spine, a catch in the voice, a faint sensation as if a distant memory of falling from a great height.

Again, the waves appear as Sagan notes, "We know we are approaching the grandest of mysteries."

The wave background again changes, this time to an ocean backdrop, focusing on calmer water, farther out from shore, and Sagan steps in front of the camera from the left. In a single glance, he projects the image of an amiable, educated person—perhaps even a professor. Nothing indicates his work as a scientist—he dons no white laboratory coat, nor does he mention his name or professional credentials. He is wearing what we'd define now as "business casual" clothing: white shirt, tan tie, tan blazer, and khaki pants under a muted orange-hued jacket. His hair—which is cut short enough so that it does not reach his shirt collar—is long enough so that it whips imperfectly in the wind.

He continues at 3:43,

> The size and age of the cosmos are beyond ordinary human understanding. Lost somewhere between immensity and eternity is our tiny planetary home, the Earth. For the first time we have the power to decide the fate of our planet and of ourselves.

At this point, the background music stops and we are left with only Sagan's voice and the sounds of nature.

The scene appears very natural and unstaged. Sagan is smiling and he's even squinting from the sun. He is not hiding under a great deal of stage makeup or hairstyling. He does not shy from showing an unflattering angle. His words are lulling but his eyes are expressive:

> For the first time, we have the power to decide the fate of our planet and ourselves. This is a time of great danger. But our species is young and curious and brave. It shows much promise. In the last few millennia, we have made the most astonishing and unexpected discoveries about the cosmos and our place within it [scene fades to seagull]. I believe our future depends powerfully on how well we understand this cosmos in which we float like a mote of dust in the morning sky.

When the scene shifts from the seagull back to Sagan at 4:35, he is standing on the edge of the cliff, his back to the camera.

He turns with a smile, and walks to join us. As he continues walking there is a sensation that we are walking together, away from the cliff's edge and the danger it implies. As he moves along, he says, "We're about to begin a journey through the cosmos . . . it's a story about us: How we achieved our present understanding of the cosmos, how the cosmos has shaped our evolution and our culture, and what our fate may be." As Sagan walks, he occasionally looks down, presumably to check his footing. The movement is natural, uncontrived.

Again, the image of Sagan is faded out to one of waves crashing, and our host indicates that our journey will focus on the "pursuit of truth." The waves are quickly replaced with sky and the camera moves downward as the narration continues to focus on the shore. It is easy to miss Sagan in the distance as he enters from the left, this time walking on the craggy beach, once again not toward the camera, but toward the shore. The image is that of a lone man on a walk, independent and unaccompanied.

At 5:50, he stops to rest and the camera focuses on him from the waist up, now leaning on a rocky protrusion, his elbow resting on the boulder, his hands clasped. He appears even to be slightly slouched as he continues to describe our coming journey. After fifteen seconds, his face appears in close-up and he says, "Some part of our being knows this is where we came from. We long to return. And we can. Because the cosmos is also within us. We're made of star-stuff . . . a way for the cosmos to know itself [a lone note of music]." Our host tells us that we will take our journey on a "Ship of the Imagination" and that we will be "unfettered by ordinary limits on speed and size." He promises that we will be able to go "anywhere in space and time."

When the scene changes to a view of Sagan, again from the waist up, leaning as before on the rock, he says,

> Perfect as a snowflake, organic as a dandelion seed [pulls a dandelion seed out from between rocks as camera zooms in on it], it will carry us to worlds of dreams and worlds of facts.

We see the dandelion's white spindly fibers swaying in the wind as Sagan grips it between his left thumb and forefinger before transferring it to his right hand. The camera returns to the image of the leaning Sagan, dandelion seed in hand, inviting us to experience the journey with him: "Come with me." He raises the seed, and he lets go. Sagan looks up into the sky, appearing to follow the now-blowing seed with his gaze.

Seven minutes into the program, the ocean setting is replaced by black space, filled not just with stars, but brilliant galaxies as well as our shining Spaceship of the Imagination, formed in the image of the dandelion seed. The galaxies in the background indicate our forward movement as they appear to be coming nearer to us. After nearly a minute, Sagan notes that we are now "halfway to the edge of the known universe," and the inside of our ship is revealed. Sagan is sitting at what appears to be a control panel set in front of a window out into space. The ship's interior is monochromatically white with a (seemingly) sharply pitched roof that is almost church-like. Sagan occupies the only chair, his back to the camera. As the camera pulls away from him, we see that the ship is long and narrow. At the opposite end, nearer to the camera, is a pentagonal "space map," which becomes clear to us at 8:30.

The camera slowly moves forward, then the scene changes to colorful, awe-inspiring galaxies close-up. Sagan then turns, gestures with both hands (palms up) toward the floor. Music plays and the floor of the ship appears to open—another window with a shape reminiscent of a keyhole. Sagan crosses his legs and leans on the arm of the chair as he continues his narration. Then, at 9:15, he turns again toward the front of the ship, clearly in control, with a map on board but no need to reference it. At 10:00, the lessons begin with an explanation of a light year, its definition and its function. The journey has begun.

Narration

The way the story is introduced and packaged plants the seeds of the program's symbolic reality. The narration is reminiscent of a bedtime story, which makes the introductory voiceovers both comforting and nonthreatening in tone or content. The language that Sagan uses is not standard scientific jargon. In fact, it isn't even language common to educational endeavors. It is dreamy and meandering, at times even seeming a bit unclear. When Sagan, leaning on the rock, says, "Perfect as a snowflake, organic as a dandelion seed" in regard to our imagined spaceship, the description seems to come from nowhere. There has been no previous reference to the perfection of space travel or in what way our travel will be "perfect." While one can eventually surmise his point (traveling using our boundaryless, limitless imagination is problem-free), this connection might not be immediately clear. The language itself is pretty and soothing, and so the audience, wrapped up in the larger concepts that Sagan describes and lulled by the quiet delivery of his narration, may not notice such slight issues of clarity.

One of Sagan's signatures as a science popularizer is his distinctive patterns of speech. While some of Sagan's pronunciations are "idiosyncratic" with "punched consonants,"[39] the manner does not extend into artificiality. In fact, the phrase for which Sagan is best known, "billions and billions," does not occur anywhere in in *Cosmos*. Writes Davidson, Johnny Carson satirized his frequent guest Sagan one night, and as a result, "Carson's comic pronunciation of 'billions' made Sagan an overnight pop reference."[40] This phrase became forever associated with Sagan's *Cosmos* persona, despite the fact that he never said it on *Cosmos*. The repetition of it, however, indicates the presence of a shared symbolic reality that stems from Sagan's host character.

Sagan is an effective popularizer "because he [is] a good rhetorician [who understands] the dramatic impact of a well-wrought phrase or gripping metaphor."[41] His words are often poetic and can at times be mistaken for something akin to a devotional. In fact, Sagan's statement, "The cosmos is all that is or ever was or ever will be," seems to reinforce this idea: it is reminiscent of a biblical description of God. However, Sagan is clearly in control of this cosmos, making him the controller of eternity and everywhere. Although the cosmos may be the realm of God, Sagan becomes the deity. These reverent first words of the program introduce audiences to the universe Sagan wants them to see, to the forum[42] he is creating. This "cosmos" becomes something more directly connected with the spiritual than with the scientific.

Sagan pushes this spiritual connection toward the literary or even poetic when he notes, in a dreamy tone, "There's tingling in the spine, a catch in the voice, a faint sensation as if a distant memory of falling from a great height. We know we are approaching the grandest of mysteries." These lines could easily be written in stanzas rather than in prose. Again, the meaning or relationship of these lines to the program's purpose is not clear. Yet, the combination of tone and artistry are inviting, indicating that this program will be more than simply "educational." The lines also indicate that this character, the program's host, is smart and sensitive. He is not challenging the audience with scientific language they may not understand. He is not delivering a boring introduction that simply tells them who he is or reviews what he's going to tell them. He touches on something more deeply human, something more interesting than a string of facts delivered with great spectacle via television. He promises the audience something not normally associated with science: beauty.

The imagery evoked through the narrator's words is meant to inspire and draw audiences into the artistic, literary qualities of *Cosmos*, much in the way a Homeric epic would. These epics "carry the unfettered reader aloft on wings sublime."[43] *Cosmos* does the same for the television viewer. The language and the visual depictions of this amazing journey, which stretches beyond the limits of the galaxy to the ends of the universe, carry the audience "aloft," as though they are on a journey with an epic hero like Odysseus. And, just as Homer's work "served as a storehouse of knowledge,"[44] so does *Cosmos* store and share the information of its contemporary scientific disciplines. Through his use of language, Sagan presents the *Cosmos* as powerful and awe-inspiring, yet controlled by universal laws of physics. At the same time he links his program with classical archetypes and story structures, which have stayed in the ken of humanity through millennia.

Sagan makes clear that this program is not "only" about science or the universe; it's not a story belonging to some "Other." It is the audience's story: "But it's also a story of our own planet and . . . it's a story about us . . . and what our fate may be." Sagan becomes a fortune teller, able to tell potential fates so we can choose our path more carefully. He evokes interest from the audience while at the same time establishing his power. He is not establishing his power *over* the audience, but instead he is establishing his power *to help* the audience.

He reiterates this message later in the introduction, making a connection not just between the audience and the story, but between the audience and the *Cosmos* itself, which now is described as part of our very essence: "Some part of our being knows this is where we came from. We long to return. And we can. Because the cosmos is also within us. We're made of star-stuff." The message again invokes spirituality. Sagan is actually describing the physical components of our being—our material make-up of matter, a grouping of atoms, recycled from the universe. However, his words indicate something more. What he describes seems to have no limitation of physicality; it is something with consciousness: "we are a way for the cosmos to know itself." We become, not just part of this godlike thing, but we become its consciousness, a key part of its structure. What Sagan describes is something more closely akin to a soul rather than animate groupings of matter. This persona is not challenging religious beliefs, as a scientist would; he is expanding science with a vision of spirituality.

Last, Sagan's use of first-person plural makes his words seem to come not from someone inaccessible and "above" (like a lofty academic), but instead they seem to come from someone who is on the same level as the audience. Although Sarah Perrault argues in *Communicating Popular Science* that using "I" and "you" makes the communication in popular science more personal,[45] it actually creates more of a separation between the speaker and the audience than using "we" does. By continually using the first person plural, Sagan puts himself *together* with his audience, creating a valuable partnership for their *shared* journey.

Throughout the entire miniseries, Sagan only rarely separates his character from the audience by saying "I." He frequently uses the terms "human species" or "we" to bring his audience closer to his message and closer to him. By saying "we," the message transforms. We, not just scientists, have explored and learned and discovered.[46] It is the human species that has done these remarkable things: the human species, of which we are all a part. Through the use of first person plural, Sagan gains credibility and further advances his character's seemingly contradictory image as a divine everyman.

Movement

Even Sagan's movement contributes to the creation of his on-screen persona. It establishes his *ethos* as an unthreatening friend or mentor who invites the audience along on a great adventure. The audience's first view of Sagan is shot from a distance at a high angle. Kress and van Leeuwen note that high angles make "the subject look small and insignificant."[47] In other words, Sagan is not presented as an intimidating presence. He is innocuous, even unimportant. In fact, he is walking toward the water, not toward the audience. He is not threatening to enter the viewers' space. The camera (and thus, the audience) moves into *his* space. When he turns to face the audience, the camera is still zoomed out, so no facial features or expressions are discernable. This man could be anyone on Earth. When the camera zooms in, Sagan addresses the audience but does not introduce himself as a scientist or even tell his name.

Sagan's first shots from far away underscore his separateness from the audience. He is very different from them, perhaps completely unknown to them. When the camera moves in closer, however, some may recognize him. The intimacy of the close shot indicates a "friendly" distance. The size of the frame isolates him and makes him vulnerable to our judgment. The flat angle depicts him as equal to the audience,[48] whereas the original high-angle shot makes him look less important. The close-up is at an "intimate distance,"[49] underscoring that he is not a threat but a friend. These shots reveal everything, even flaws. The audience is able to see directly into his eyes. Sagan is unable to "hide" anything. His entire countenance is up front and center. Nothing is concealed from view. Baring Sagan's face and flaws to the audience indicates a lack of ulterior motives. This visual representation provides an "imaginary relation" between the audience and the character depicted.[50] The audience may see characters shown in close-up as "friends and neighbors,"[51] even though they are strangers.

These shots occur often throughout the thirteen episodes. In the 1980s, color television sets were the norm but they still were not as large as modern sets. They also did not show pictures in high definition, so close-ups may not have been quite as overbearing to the program's contemporary audience as they may seem today. Of course, some critics and critical viewers noticed the close-up shots and consigned them to Sagan's arrogance. However, these close angles serve a rhetorical purpose in "closing the distance" between the shepherd and his flock. The close-ups reinforce the fantasy theme of Sagan as friend.

Because of his movement toward a neutral position, Sagan is unintimidating. The audience, via the camera's perspective, joins him and enters his space: he does not invade our space. Characters who move forcefully can be seen as "commanding,"[52] abrasive, or aggressive. Sagan's relaxed walk toward the cliff and then toward the water eliminates this potential interpretation of his character as being threatening. When Sagan stands at the edge of the cliff, his back to the camera, he turns around so he can talk to the audience. As he turns, he smiles, as if greeting someone familiar to him. He does not turn quickly, with anticipation: he turns slowly, with no indication of feeling threatened or uncomfortable. Then, as he walks toward the camera, it begins moving with him. Sagan's face is shown more closely, but the camera is pulling backward as he moves. The sensation is that the audience is now walking with him, a companion. Because Sagan must look down and check his footing, a sense of realism is brought to the scene. It is a natural (and intelligent) movement for someone walking on a rocky cliff (conversing with a companion) to watch where he is stepping. He is not staring into the camera from a short distance; he releases his "eye contact" with the camera, increasing the realism of the scene. The audience has entered his space, and we have joined his walk. This joining indicates the audience's power to choose whether or not to go along on his journey. Sagan's character presents an image of welcome helpfulness, not of demand.

Setting

Beginning the *Cosmos* journey at the edge of the ocean symbolizes the evolution of humans from sea creatures. It was a gigantic leap between the water and the land, just as the leap between the Earth and the stars seems enormous to us now. The rhetorical fantasy of evolutionary destiny takes root on Sagan's craggy beach. The crashing waves that open the program also demonstrate the power of nature. Of course, the natural beauty of the location would have interest for the audience. However, it also allows Sagan to demonstrate his own willingness to be vulnerable and to emphasize that this journey is being offered via invitation, not decree. The often and quickly shifting images on screen contrast with the slowly delivered monologue, adding interest. When Sagan is shown in his first close-up with his hair whipping in the wind, he doesn't fuss with it. He doesn't wear a hat and he clearly doesn't have a crew off-camera blocking the wind. He shows that he is no *prima donna* by allowing himself to be filmed in somewhat unflattering conditions. Because he must check his footing while he walks, the audience is assured that he really is walking on this uneven terrain: there is no hidden "sidewalk" which he enjoys but which is unseen by them. Sagan is presenting his character as though he were just a person out by the ocean. His lack of fussiness about his appearance, his willingness to squint his eyes rather than hiding them behind sunglasses is an exposure of his "true" self. As a sign of his sincerity and trustworthiness, he does not appear to hide behind makeup or other falsifications of television.

When the camera comes (for the first time) from behind Sagan, he is standing on the edge of the cliff. The body language places him in a vulnerable position. He does not need to be in control of the audience's actions and he trusts the audience with his own safety. This vulnerability indicates that his character will confer onto audiences the same thing that he asks of them: trust. He needs them to believe what he has to say, to accept it without a lot of doubt or hesitation. The territory into which he intends to travel is controversial at times, but he must have the chance to explain his stance on those issues before other prejudices or preconceptions interfere. He wants the audience to trust him enough that they will stand at the edge of the cosmos's cliff and learn what they may have previously perceived to be "dangerous" to their beliefs or identities.

The choice of a dandelion to represent a spaceship is also an interesting and unusual one. Normally, spaceships are depicted with many mechanisms for defense or attack. But not Sagan's. His dandelion is innocuous, beautiful, and functional. While any ship that can travel through space must be equipped with impressive technology to power it and to preserve the life inside of it, this ship is related to the common dandelion. Although powerful by necessity, the ship's lack of preparation for attack or defense signals an unthreatening agenda. Therefore, the "captain" of this ship, our host, must also be unthreatening in his motives. In addition, a ripe dandelion seed spreads life everywhere. Rather than being a symbol of danger, this ship is being associated with spreading life.

The spaceship's stark white and church-like interior is again reminiscent of something religious. Because Sagan has told us that we are "star-stuff," we already know that we are woven into the fabric of the *Cosmos*. The lack of color within the ship may be reminiscent of popular images of "heaven," but it is also a nod to the "colorlessness" of a black and white universe. The map at the back of the ship, far behind the location from which, presumably, the ship would be steered, provides a visual anchor to the scene. Sagan clearly knows this map well enough that it can be located two or three meters behind the control panel without issue. His knowledge of that map is so thorough and reliable that he has little need to reference it, nor does he need a copilot. Only one seat is available, and Sagan has it. He clearly steers the ship, guiding us anywhere in space or time. We trust him and he gains credibility through this display of competence.

Sagan himself is not presented in white nor in a laboratory coat. Such a depiction would at least *differentiate* him, and potentially *elevate* him, "above" his audience. Sagan wants to present his host's persona as an ordinary person—one who is very knowledgeable, of course, but one who fits in with the average person. He wears a tie, which distinguishes him somewhat from other people as a signal of authority. However, it is a textured tie rather than a smooth or expensive silk tie. His corduroy blazer also adds texture to his dress, eliminating the possibility of appearing too polished. In fact, this blazer is almost a prototypical "professor" blazer. So, even though Sagan does not present his credentials nor even say his own name, he creates through his dress an amiable professor persona, distancing himself from an identity as a Scientist. The character presented is a reliable guide, peaceful, curious, inviting, smart, calm, and friendly. He is divinely powerful, accommodating, and self-less. Through Sagan, the audience will discover the great sights and astonishing truths he has promised.

SCIENTIFIC *ETHOS* AND THE SCIENCE POPULARIZER

Creating a fantasy theme within a scientific program isn't easy—and neither is it without professional impact. Because science as a field relies so heavily on the norms and guidelines it has established, someone who presents science outside of those barriers may represent a challenge to the field and a danger to its equilibrium. This potential threat is tied to the cumulative scientific *ethos*. According to Francis Bacon, reputation built over time influences the audience's perception of the rhetor's performance, because "his popular reputation is always an important factor in the audience's estimation of his character."[53] Speakers do not start each presentation with a clean slate unless they are wholly unknown to the audience. Reputation-based *ethos* can either benefit or harm a rhetor. An individual with an established, positive reputation will face less of a challenge in convincing the audience to listen to the message delivered. An individual with a poor reputation, however, will have additional challenges in persuading an automatically skeptical audience.

Speakers may also tap into a "cumulative" *ethos* through association with a group that is already known to the public. By claiming an association with a known political, social, economic, or intellectual group, the speaker is putting on a mantle of the values and actions of that group. For both Merton and Halloran, the scientific *ethos* is collective, shared by all scientists[54] and not the result of individual reputations. The invocation "science" endows speakers with great inherent power.[55] Because science is generally considered a barometer of "reality," every contact with a scientist no matter how mundane or innocuous could result in some type of persuasion.[56] Beginning from a place of great power lends great responsibility—and great culpability—to those who enjoy that advantage. To preserve it, science has set up mechanisms within the field to protect the positive aspects of science's cumulative reputation.

Gatekeeping mechanisms like peer review protect science's rhetorical advantage.[57] Scientists are trained to conform their communication practices with those of their field to be accepted as a member of the scientific community; this training teaches them how to convince peer reviewers or journal editors that their work is important and that it has been done correctly.[58] To reach a wide audience, scientists must first persuade these peer gatekeepers of their worthiness. Overington charges that scientists who do not follow the communication process into which they are apprenticed can be ostracized or even discredited within their profession. In addition to this potential for blacklisting or snubbing, individual scientists' reputations are based, at least in part, on their ability or willingness to conform to standards.[59] Refusal or the inability to comply with these communication norms may result in a jeopardized career. Merton describes the science communication renegades (and failures) as "caste-inferiors" who have demonstrated they don't measure up to "caste-standards"—and that these people must have "their contributions . . . systematically devalued."[60]

Such serious consequences could be tied to the way science builds on previous research. Scientists who don't follow one protocol may not follow others, so perhaps it is easy to imagine that their scientific work might have rejected other established norms as well. Rogue scientists whose processes can be questioned cannot be relied upon to produce dependable work. If their work slips by, then future scientists who attempt to build on it could suffer humiliating failure. Extrapolated to an extreme, the collective scientific *ethos* could be thrown into doubt. The "discourse community"[61] of science may see the rejection of standards as a threat to the scientific collective *ethos*.

Scientists who threaten this "currency" of the field must then be separated from the group.

These mechanisms establish communication norms within science and between scientists. However, popular science cannot share the same gate-keeping mechanisms with orthodox Science. Whether popularizations are in the form of articles in the mainstream press or are television programs, they are destined for a broader audience than any academic article can hope to have, regardless of importance. Sometimes, these popularizations are written by people with limited scientific backgrounds; other times they are written by credentialed scientists. However, those who accommodate or popularize scientific information open themselves to potential criticism from both lay audiences (e.g., criticisms regarding lack of clarity or appeal) and by scientists (e.g., claims of inaccuracy or oversimplification). While Merton writes about the "the institutional structure of science,"[62] it should be acknowledged that popular science's "institutional structure" should lie, to some degree, with the medium through which it is distributed.

Because the institutional structure for popular science does not match that of orthodox science, Carl Sagan was part of the "science collective" while also being a science popularizer, which meant he regularly communicated outside of the field's established channels. His popularizations did not receive the type of field endorsement that peer reviewed articles would have. Instead, this scientific information was transferred outside the boundaries and control of the field. Just as Merton noted, scientists who do so will be "ostracized." Although some may argue that "ostracized" is an overstatement, Sagan was no exception.

A boundary-breaker and bar-setter for science popularizers, Sagan is at times disparaged and disrespected within his own profession. His *ethos* as a scientist has been questioned—and it was even dismissed when he was refused entry to the National Academy of Sciences (NAS) in 1992. Deliberations were held confidential, but debate was reportedly vigorous the year his name was considered.[63] That year, the vote was very close but only fifty-nine instead of the usual sixty scientists were admitted for membership in this elite group.[64] Denying Sagan entry indicated that fellow scientists didn't deem Sagan "worthy" of the honor. They questioned his contribution to the field of science, despite the fact that his enormous public appeal resulted in conference talks about science that needed to be booked in stadiums, not classrooms or even auditoriums.[65] Even though he later received the NAS' Public Welfare Medal, the message the voting body sent was clear: Sagan was being denied the achievement and honor of being counted among the membership of elite scientists. He was effectively marginalized, a "caste inferior" in the field of orthodox science.

Reportedly, the main criteria for rejecting him was that his research and discoveries were not significant enough, and that admitting Sagan would "open a floodgate to people whose science isn't spectacular but [who] have other credentials."[66] These "other credentials" could indicate that the academy thought Sagan had blurred the line between science and popular science, so his work as a whole could not be easily categorized. It is hard to say if the objections were directly related to the creation of "persona," an illusion, when science is based on facts. It could have been based on Sagan's inability or unwillingness to follow the established "rules" of scientific communication. Potentially, it was the threat to the established systems of science posed by mass popularization—such as *Cosmos*—that spurred their rejection of him. If the sciences are really "knowledge making cooperatives," as Polanyi[67] claimed, which rely on their "integrity as discourse communities,"[68] then ostracizing someone who threatened whatever controls were set in place to protect the scientific discourse community may have made sense to the collective.

The rejection may also be related to the power a great popularizer can wield and how that popularizer, in some ways, receives "ownership" of the science in the public's eye. Audiences gave Watson and Crick credit for the discovery of DNA, but these two scientists didn't work alone. However, it was their article that announced the finding (in the most accessible way)—and they were thereafter given credit by the press and general public.[69] Sagan didn't need the "cooperative's" consent to create popularizations and so the potential existed for him to inadvertently "get credit" for science that wasn't his work. Whether some members of the NAS voting body were worried about Sagan usurping credit, after his success as a popularizer became obvious, his scientific colleagues viewed him as more of a "showman" than as a scientist.[70] As a result, he "could no longer be regarded as a serious scientist."[71]

This perception is patently unfair since it may not have been Sagan's lack of contribution or "inability" as a scientist that led to his rejection. In fact, Sagan possessed a Ph.D. in astronomy and astrophysics, had a *curriculum vitae* that was 265 pages long, worked with NASA on Mars missions and the Voyager probe, and published an average of one "scientific peer-reviewed paper per month"[72] from the start of his career in 1957 until his early death in 1996. He may have published more popular science articles after *Cosmos* than he did beforehand, but his rate of scholarly paper publication did not decrease.[73] He maintained his connection to orthodox science and discovery even as he expanded public knowledge of, interest in, and funding for science. His contribution to science was different from that of other scientists but arguably just as valuable (if not more so) than most. In his case, the ability to tap into a cumulative *ethos* of a group helped him to have credibility with audiences but ended in a professional marginalization of sorts that was inflicted because of a lack of understanding or acceptance of his methods—even though they were enormously successful—because these methods did not follow with what is traditionally expected of a scientist.

NOTES

1. Daniel J. Kevles, *The Physicists: The History of a Scientific Community in Modern America* (New York: Alfred A. Knopf, 1978), 397.
2. George Gerbner, "Science and Scientists on the TV Screen," in *AAAS Annual Meeting* (Toronto, 1981), 5.
3. Ernest G. Bormann, "Fantasy and Rhetorical Vision: The Rhetorical Criticism of Social Reality," *Quarterly Journal of Speech* 58, no. 4 (1972): 398, doi:10.1080/00335637209383138.
4. Lynda Walsh, *Scientists as Prophets* (New York: Oxford University Press, 2013), 146.
5. Ibid.
6. Ibid., 149.
7. Ibid.
8. Bormann, "Fantasy and Rhetorical Vision: The Rhetorical Criticism of Social Reality," 401.
9. Walsh, *Scientists as Prophets.*
10. James S. Baumlin, "Introduction," in *Ethos: New Essays in Rhetorical and Critical Theory*, ed. James S. Baumlin and Tita French Baumlin (Dallas: Southern Methodist University Press, 1994), xv.
11. Paul Newell Campbell, "The Personae of Scientific Discourse," *Quarterly Journal of Speech* 61, no. 4 (1975): 391–405.
12. George Yoos, "A Revision of the Concept of Ethical Appeal," *Philosophy and Rhetoric* 12, no. 1 (1979): 47.
13. S. Michael Halloran, "Aristotle's Concept of Ethos, or If Not His Somebody Else's," *Rhetoric Review* 1, no. 1 (1982): 60, http://www.jstor.org/stable/465559.
14. James S. Baumlin and Tita French Baumlin, "On the Psychology of the Pisteis:Mapping the Terrains of Mind and Rhetoric," in *Ethos: New Essays in Rhetorical and Critical Theory*, ed. James S. Baumlin and Tita French Baumlin (Dallas: Southern Methodist University Press, 1994), 99.
15. Robert K. Merton, *The Sociology of Science: Theoretical and Empirical Investigations* (Chicago: University of Chicago Press, 1973).
16. Jeffrey W. Kirsch, "On a Strategy for Using the Electronic Media to Improve the Public Understanding of Science and Technology," *Science, Technology, and Human Values* 4, no. 27 (1979): 54.
17. Walter B. Weimer, "Science as a Rhetorical Transaction: Toward a Nonjustificational Concept of Rhetoric," *Philosophy and Rhetoric* 10, no. 1 (1977): 22.
18. S. Michael Halloran, "The Birth of Molecular Biology," *Rhetoric Review* 3, no. 1 (1984): 74, http://www.jstor.org/stable/465734.
19. Ibid.

20. Ibid., 72.
21. Keay Davidson, *Carl Sagan: A Life* (New York: John Wiley, 1999), 323.
22. Ibid., 190.
23. Ibid., 323.
24. Ibid., 325.
25. Ibid., 52.
26. Ibid., 333.
27. Ibid., 332.
28. Marilynn Preston, "Looking for Miracles: Seton's Life, Sagan's Last Stand," *The Chicago Tribune*, December 19, 1980, A19.
29. William J. O'Malley, "Carl Sagan's Gospel of Scientism," *America Magazine*, February 7, 1981, http://americamagazine.org/issue/100/carl-sagans-gospel-scientism.
30. Kristin McMurran, "His Cosmos a Huge Success, Carl Sagan Turns Back to Science and Saturn's Rings," *People*, December 1980, http://www.people.com/people/archive/article/0,,20078104,00.html.
31. Preston, "Looking for Miracles: Seton's Life, Sagan's Last Stand," A19.
32. Davidson, *Carl Sagan: A Life*, 332.
33. Frederick Golden, "The Cosmic Explorer," *Time*, October 1980, 64.
34. Davidson, *Carl Sagan: A Life*, 331.
35. McMurran, "His Cosmos a Huge Success, Carl Sagan Turns Back to Science and Saturn's Rings."
36. "Shores of the Cosmic Ocean"
37. "The Backbone of Night"
38. Adrian Malone, "Shores of the Cosmic Ocean," *Cosmos: A Personal Voyage* (Studio City, CA: Cosmos Studios, 1980).
39. Davidson, *Carl Sagan: A Life*, 332.
40. Ibid., 333.
41. David J. Tietge, *Rational Rhetoric: The Role of Science in Popular Discourse* (West Lafayette, IN: Parlor Press, 2008), 120.
42. The idea of forum is explored in more depth in chapter 4, "The Forum of Television."
43. Samuel E. Bassett, "Homeric Criticism," *The Sewanee Review* 30, no. 4 (1922): 393.
44. Ibid., 395–96.
45. Sarah Tinker Perrault, *Communicating Popular Science* (New York: Palgrave Macmillan, 2013), 106.
46. Thomas M. Lessl, "The Priestly Voice," *Quarterly Journal of Speech* 75, no. 2 (1989): 191.
47. Gunther Kress and Theo van Leeuwen, *Reading Images: The Grammar of Visual Design* (New York: Routledge, 1996), 146.
48. Ibid.
49. Ibid., 131.
50. Ibid., 132.
51. Ibid.
52. Kirsten Marthe Lentz, "Quality versus Relevance: Feminism, Race, and the Politics of the Sign in 1970s Television," *Camera Obscura* 15, no. 1 (2000): 71.

53. Ibid.
54. S. Michael Halloran, "Aristotle's Concept of Ethos, or If Not His Somebody Else's," *Rhetoric Review* 1, no. 1 (1982): 62, http://www.jstor.org/stable/465559.
55. Philip C. Wander and Dennis Jaehne, "Prospects for a 'Rhetoric of Science'" 14, no. 2/3 (2000): 214.
56. Walter B. Weimer, "Science as a Rhetorical Transaction: Toward a Nonjustificational Concept of Rhetoric," *Philosophy and Rhetoric* 10, no. 1 (1977): 14.
57. Philip C. Wander, "The Rhetoric of Science," *Western Speech Communication* 40, no. 4 (1976): 230.
58. Michael A. Overington, "The Scientific Community as Audience: Toward a Rhetorical Analysis of Science," *Philosophy and Rhetoric* 10, no. 3 (1977): 143–164.
59. Robert K. Merton, *The Sociology of Science: Theoretical and Empirical Investigations* (Chicago: University of Chicago Press, 1973), 272.
60. Ibid., 227.
61. R. Allen Harris, "Rhetoric of Science," *College English* 53, no. 3 (1991): 283, http://www.jstor.org/stable/378102.
62. Merton, *The Sociology of Science: Theoretical and Empirical Investigations*, 269.
63. Faye Flam, "What Should It Take to Join Science's Most Exclusive Club?," *Science* 256, no. 5059 (1992): 960–61.
64. Ibid.
65. Duncan Steel, "Carl Sagan: Practitioner, Popularizer and Proponent of Science," *Contemporary Physics* 42, no. 4 (2001): 248.
66. Flam, "What Should It Take to Join Science's Most Exclusive Club?," 960.
67. Michael Polanyi, *Personal Knowledge: Towards a Post-Critical Philosophy*, Harper Tor (New York: Harper & Row, 1964).
68. Harris, "Rhetoric of Science," 283.
69. S. Michael Halloran, "The Birth of Molecular Biology," *Rhetoric Review* 3, no. 1 (1984): 70–83, http://www.jstor.org/stable/465734.
70. Keay Davidson, *Carl Sagan: A Life* (New York: John Wiley, 1999), 331.
71. Ibid.
72. Michael B. Shermer, "The View of Science: Stephen Jay Gould as Historian of Science and Scientific Historian, Popular Scientist, and Scientific Popularizer," *Social Studies of Science* 32, no. 4 (2002): 493, http://www.jstor.org/stable/3183085.
73. Ibid.

Chapter 4

The Forum of Television

The popularity of *Cosmos* stems from its ability to "transport" its audience. Viewers are willing to join Sagan as their guide on the journey because they trust him. He establishes his credibility and shows that he is not a threat. The next step for the program is to build the "vehicle" for transportation. From this vehicle, depictions of space and the earth merge to create the program's major setting (fantasy) themes. These in turn contribute to building the program's rhetorical vision. Bormann clarifies the term "rhetorical vision" to mean "the shared symbolic ground exhibited" by those who participate in its formation and the culture it produces.[1] The vision (or symbolic reality) is created by "integrated symbol systems"[2] that trigger a common belief or reaction from a group of people. Symbolic realities are often reconstructed from fantasy chains gathered from multiple sources (political rallies, television ads, billboards), with all contributing to a single rhetorical vision.[3] The major setting themes of *Cosmos* reflect its rhetorical vision while also allowing it to build on the symbolic realities of science fiction cinema. *Cosmos*' setting themes exploit science fiction fan culture to make itself more appealing or interesting to viewers, even as it works to build its own rhetorical vision.

In *Cosmos*, the audience is able to "join" in the journey that Sagan undertakes in a sort of "virtual reality" experience. He even christens his ship "Spaceship of the Imagination," making plain its grounding in dream and fantasy. The audience becomes part of the program's adventures, and they join into the symbolic reality created through the program. Because the audience is on the journey with Sagan, they become witness to the events depicted on the screen. They come to share a common vision with Sagan, making them more open to persuasion, to see things as Sagan does. For example, in episode 13, "Who Speaks for the Earth,"[4] Sagan poses as a space traveler who comes upon a world that is "clearly inhabited, a world I had visited before." This world has all the signs of an advanced civilization, but while he's watching it, he says, "suddenly, darkness. Total and absolute." His computer helps him to determine what has happened. It flashes pictures of a variety of planets (including Earth-like blue planets) and then settles on the half-lit, suddenly dark world he earlier referenced. As Sagan reads through the planet's statistics, he says, "Their biology was different from ours. High technology . . . . They must have known they were in trouble."

He begins to feel homesick for Earth, so, he begins to listen to television broadcasts from "home." These broadcasts feature quickly changing news announcements that indicate some type of trouble—then they are quickly cut off. No sound replaces them until Sagan again says: "Then suddenly silence. Total and absolute." Sagan has not yet told the audience to equate the alien world and what happened there with Earth's sudden silence, but the similarities are clear.

Sagan also does not say what happened to Earth, why the transmission suddenly cut off, but he declares that the destruction would have been "at our hands." He provides a list of ways that the earth "could have been" destroyed, including the "ravaged land," "poisoned air," "changed climate," plague and nuclear war. Sagan pulls up information pictures, statistics, and numerical identification on his computer, just as he had done with the alien planet earlier. The age of the earth is shown to be the same as the alien world.

In the end, Sagan determines that it is nuclear war that destroyed the earth. He uses emotionally triggering words, describing the death of humanity as meaning "never again a love, or a child . . . no more songs from the earth." Then the camera focuses on his profile, and he bows his head in mourning. After he ponders the weaknesses of human life (and "the reptilian voice within us"), he asserts that, "We accepted the products of science; we rejected its methods." At the end of the monologue, he laments that there would "never be another human, not on a billion worlds." The image of Sagan's back while gazing out the front window of his spaceship at Earth combines with the image of a mushroom cloud. Sagan's image fades as additional explosions and clouds are presented.

The framing of the two planets, both destroyed, builds a setting theme surrounding the dangers of nuclear war. Sagan builds on already-existing symbolic reality about mushroom clouds as immensely dangerous, terrifying weapons to persuade his audience of their power to wipe out not just a target, but all life on Earth. It took Sagan and TTAPS' term "nuclear winter,"[5] coined three years after *Cosmos* first aired, to get the sort of resonance he wanted in regard to nuclear war. However, the attempt to communicate this same sentiment of impending apocalypse is part of the rhetorical vision of *Cosmos*. Sagan does not tell the audience directly that they should resist nuclear proliferation; he shows the evidence of his argument and lets them draw the conclusions.

In this way, *Cosmos* uses a particular kind of visual framing based in the idea of *ethos* as haunt or forum. This framing can be used as a way to put an audience in an intellectual or emotional "space" that prepares them for persuasion. The concept of a rhetorical forum stretches back to Aristotle, even though Greek word *ήθος* is traditionally transliterated to *ethos* and defined as the credibility of the speaker. In actuality, its full definition includes "forum," "haunt," or "dwelling place." While both meanings of *ήθος* can be understood in the original Greek, the second important aspect of the term is often forgotten. The traditional definition of *ήθος* was explored in depth in chapter 3, "*Ethos* and the Popular Science Persona," so this chapter will concentrate on the "other side" of *ήθος*: the forum.

This investigation will extend the idea of forum to include frames, built through sensory manipulation, that unite audiences in a single vision. This unification of vision reinforces the effectiveness of setting (fantasy) themes. Television can build such a framework through visual (and even auditory) stimuli. Although audiences are made of an amalgamation of disparate individuals who can be separated by space and even time, this visual framing can create a common lens through which they all can internalize the rhetorical vision of the rhetor. The visual elements of *Cosmos* connect it to popular entertainment but they also provide the symbolic cues (what Bormann equated to "inside jokes"[6]) that trigger its fans' communal connection to the program.

ĒTHOS AS FORUM

The idea of *ethos* as forum has been explored a bit by rhetoricians in the last few decades; as Charles Chamberlain, S. Michael Halloran, and Dale Sullivan have pointed out, the concept of a forum or dwelling place is an important rhetorical tool. They all use different terminology to describe the concept, but I will adopt Chamberlain's transliteration of *ēthos* to help differentiate between the idea of forum and the idea of credibility (usually transliterated as *ethos*). This distinction, although slight, provides at least some visual difference between the terms.

The appeal of *ēthos* in many ways is launched through the senses. George Yoos describes it as a sensory-based "emotive appeal"[7] and acknowledges that its "rhetorical qualities . . . are analogous in many respects to aesthetic qualities."[8] This aesthetic angle to *ήθος*, according to Thomas E. Cort, was potentially lost in an "etymological" misunderstanding or "mix-up."[9] The literal translation of *ēthos* has been described by S. Michael Halloran as being about physical space. He writes, "The most concrete meaning given for the term in the Greek lexicon is 'a habitual gathering place.'"[10] In other words, *ēthos*, as Halloran defines it, is about a regular, understood space wherein the community can meet under a unifying condition or purpose of some sort.[11] The idea of "sharing space" is reinforced by Dale Sullivan when he describes a unification of the audience, transforming them from individuals to parts of a whole. However, Sullivan uses the term "consubstantiality" to describe an experience that extends beyond the physical to the spiritual.[12] This definition is more in line with the current interpretation of *ēthos* as an intellectual or emotional space, which primes an audience for persuasion. Consubstantiality can also denote a shared experience.

Sullivan's definition provides some fitting terminology for discussing the role that *ēthos* plays in persuasion, but Charles Chamberlain's work also provides illumination. Chamberlain explains that Homer defined the space of *ēthos* not as a location where people "habitually meet," but rather as a space that is used to habituate creatures to new routines. He uses the example of stabling horses to illustrate his meaning: "The point of the manger is that the humans hope to change the horse's habits by using hunger to make it regard the manger as home."[13] Hunger is a matter of physical perception—a feeling of sorts. It can trigger anxiety or other emotions, depending upon the individual's past experiences and present circumstances.[14] Being driven by an instinctual habit of returning to the source of food, a horse can be "programmed" to see the place (the manger) as a positive location. Because of the difficulty in "rehabitualizing" the creature, a correlation is drawn with persuasion. Even though it is difficult to rehabituate a horse (who may still feel the pull of the pasture), it is not impossible.[15]

Using a physical space to change an inward belief or reaction is the connection Chamberlain is making, but he does not imply that it *must* be a physical space where this change occurs. Herein lies the power of *ēthos*: the difference between a physical space for meeting and a *virtual place* for creating mental change (or making an emotional impact). If the space is "merely" physical, the rhetorical usefulness of *ēthos* might be limiting.[16] But, because this space can also be constructed as a mental or even emotional state, the rhetor can effectively shape the psyche and perceptions of the audience to elicit the response desired.

In a more recent exploration of the concept of forum, Michael J. Hyde asserts that "the *ethos* of rhetoric" focuses on communication that "transforms space and time" into settings that allow people to consider shared knowledge or interests.[17] These places influence and educate audiences, attracting them through being "aesthetically, psychologically, socially, and perhaps theologically instructive."[18] Hyde notes that Aristotelian *ethos* focuses not as much on behavior as on the power to "inspire trust in an audience."[19] According to Hyde, *ēthos* depends on the credibility of the speaker in that the audience must trust the speaker before entering "the right frame of mind so that their emotional state during the oration and subsequent deliberations are advantageous to the speaker's persuasive intent."[20] This "frame of mind" is critical to the concept of ἦθος as forum. It is the intellectual and even emotional space in which an audience becomes ripe for persuasion.[21] While *kairos* captures the spirit and interest of the time, the credibility of the speaker (*ethos*) must be established before the audience will be mentally or emotionally receptive to the forum of persuasion (*ēthos*).

This space connects the audience emotionally to the speaker's argument and provides clues for the "appropriate" (or the rhetor-desired) audience reactions. The frame created by the speaker is "entered into" by the audience, and it is from this space that the audience experiences the arguments and information presented. When the audience enters the speaker-created forum, the speaker can then control the viewpoint from which they experience that information. Because the audience does not view the information presented through their own individual "frames" but instead through the frame or forum created by the speaker, attempts at persuasion can be highly effective.

Getting the audience to be empathetic to a situation, or getting them to feel as though they are *participating* in the situation, is key to persuading them. Hyde's explanation of the haunt or forum focuses on internal processes within the audience. He notes that these dwelling places can be inviting "symbolic constructions," but the forum must make the audience be comfortable or "at home" while they are considering the rhetor's argument.[22] Naturally, television allows audiences to be *literally* "at home" when they receive popular science programs like *Cosmos*. It is a highly visual medium and fully capable of triggering internal processes, such as those described as *ēthos* or forum. Therefore, television has great power and opportunity to heighten the effectiveness of arguments (scientific, historical, etc.) presented. The audience receives evidence through the lens that the program chooses, rather than just their own. Within the context of television, the forum manipulates the senses and emotions of the audience. The setting themes created thereby contribute toward the series' overarching rhetorical vision.

THE RHETORIC OF PLACE AND TELEVISION

A virtual "place" created by television can have the same rhetorical benefits as physical space. Danielle Endres and Samantha Senda-Cook assert that it is only "places themselves—not discourse about places" that create opportunities for persuasion.[23] They argue that simply speaking about the place will not move the audience to action,[24] but in truth, a reference to a physical location can elicit the same emotions and may even encourage more open expression of those emotions. Television functions in the three ways that Endres and Senda-Cook outline as the way place acts rhetorically. The first way is by "build[ing] on a *pre-existing meaning of a place* to help make [a] point."[25] The second is by "*temporarily reconstruct[ing] the meaning* . . . of a particular place."[26] The third is by being able to morph and change through "*repeated reconstructions over time*."[27]

Pre-Existing Memory

First, television can easily build a "pre-existing memory of place" by displaying images that are likely to trigger memories for the expected (or desired) audience. The earlier example of *Star Wars* imagery being used in *Cosmos* can also serve as an example here. The familiarity of a beloved environment can increase audience comfort. Without these referential touchstones, the universe in *Cosmos* may have seemed foreign, cold, or uninteresting to contemporary audiences. According to Marcel Chotkowski La Follette, at least until *Cosmos* first aired in September of 1980, presentations of science on television emphasized the "scarier" or more sensationalist possibilities of the field.[28] Television sometimes supported or even at times emphasized a vision of space as being an "exotic and dangerous" location.[29] So, it was logical for *Cosmos* to make purposeful, overt connections with previous science fantasy entertainment like *Star Wars* (rather than previous science programs) to "recast" its image. Sagan rejected one sort of pre-existing memory to emphasize another.

Temporary Re-Experience

Second, television reconstructs a "temporary" meaning or a temporary place. For example, a program may be thirty minutes long and afterward the viewer may step away and perhaps never return. However, when programs go into syndication or when they are released on video, they remain essentially unchanged from the original. Additional elements may be packaged in a "bundle" if purchased (DVD extras), but the episodes are normally left, even with their flaws, preserved for history. Theatrical releases may not be as unchanging as they may go through processes of restoration, colorization, modification to add new effects like 3-D, or revision for ideological reasons.[30] However, television programs, and popular science programs in particular, are not usually targets for editing of this kind and then rereleased to the public. *Cosmos* was edited more than once after its initial release and some updated commentary featuring Sagan was added at the end of episodes. However, the bulk of the original remained untouched and changes were mostly unnoticed by audiences.

These "locations" created on the screen remain dependably the same but the audience only visits them *temporarily*. In some ways, the "timelessness" of this space (as Dale Sullivan[31] describes it) creates an additional trust between the speaker and the audience. In the case of television popular science, the arena will stay the same even if audiences revisit it in twenty or thirty years. The space created for the shared learning or experience is held as sacred, inviolable. This sense of the sacred is why, when attempts are made to change the established shared space, audiences cry out in protest.

For example, the updates that George Lucas made to the original *Star Wars* trilogy have spawned much debate and even "fan" hatred. He committed the sin of violating the shared space of those films, thereby nullifying the consubstantiality and breaking the "enfolding" aspect of the exchange. In the case of the 2014 *Cosmos* soft reboot, fans of the original could have ended up being less receptive to it than they had been to the original—and that may be one reason why it spent so much time honoring Sagan by tipping its "hat" to him as well as "passing the baton" between Sagan and Tyson. Regardless, these existing fans are unlikely to have been the primary target audience for *Cosmos* 2014. Were Sagan still alive or had he passed away recently and not nearly twenty years before, reception of Tyson as the rebooted program's host may have been different.[32]

Repeated Reconstructions

The third way in which television can act rhetorically by creating "place" is by being able to repeat and reconstruct the environment over time. *Cosmos* ran over a thirteen-week period, one hour per week. The audiences that came back and watched the entire series saw the repeated framework; they experienced familiarity of the forum over time. While the "locations" depicted may not have been the same every night, the approach to building and introducing them stayed consistent throughout the entire series run.

Beyond that type of comforting predictability, repetition is also possible with syndication and sales of the program on video or DVD formats. This consistency helps fans learn the environment. Fan homages to Sagan and *Cosmos* on YouTube testify to the tangibility and importance of this space to viewers who have repeatedly visited it. The space is reconstructed—in full—with repeated viewings. It is also reconstructed—with new elements—through parody and fan homages.

One of the most impressive visual elements created for *Cosmos* was the Library of Alexandria. The visual effect is so well done that it may not have been obvious at the time that it is computer animation and not a location shoot or full-sized set piece. While the library is referenced in more than one episode, the final episode, "Who Speaks for the Earth," provides most detailed look at it. The illusion of "place" as a physically distinct location is strong in this scene, even though it is essentially an artist's imagining. According to Perelman and Olbrechts-Tyteca, "An illustration designed to create presence will sometimes have to be developed with a wealth of concrete and vivid detail."[33] In this instance, not only are the computer renderings impressive in their detail but when Sagan walks across the library to "meet" the audience, a shadow falls across his face. The shadow is not present when he speaks of the Library's great achievements, but when Sagan delivers the lines that could be most closely applied to the politics of *Cosmos*' time, the shadow appears.

He stops on his mark in just the right-sized piece of light so that his face is well-lit, even though shadows still show across his body and when he moves. Speaking directly about ancient Alexandria but implicitly about the state of science in the United States, Sagan says:

> The vast population of this city had not the vaguest notion of the great discoveries being made within these walls. How could they? The new findings were not explained or popularized. The progress made here benefitted them little. Science was not part of their lives. [*Shadows move in*]. [Scientific discoveries] mainly were applied to the perfection of weapons, to the encouragement of superstition, to the amusement of kings . . . . The great intellectual achievements of antiquity had few practical applications. Science never captured the imagination of the multitude . . . . So when, at long last, the mob came to burn the place down, there was nobody to stop them.

Because this is the final hour of the series, Sagan is less veiled with his arguments. The detail of this final recreation of the library emphasizes the importance of the message Sagan delivers in the segment: science must be understood and valued by everyone if a society is to advance. His "presence" in the library makes the argument more immediate. It also creates a symbolic connection between this ancient civilization and contemporary America. Standing on the spot of this intellectual and social tragedy gives Sagan's argument about the importance of science (and public understanding of it) more veracity.

Although Sagan is not *really* standing in the Library of Alexandria, his message receives a rhetorical "boost" like it would if he were really there. The audience, too, is sitting at home watching television. Yet, the recreation of this place (*ēthos*), coupled with the implications of Sagan's cautionary tale, is likely to move them more than if he were shown in an empty set with a blank background. This "place" is created visually through television, and *Cosmos* used that recreation to increase its rhetorical impact.

CREATION OF FORUM THROUGH IMAGES

Kress and van Leeuwen argue that "images communicate meaning."[34] After all, letters and words are a collection of "signs" that can be used for persuasion, so it makes sense that visual stimuli can operate in much the same way.[35] Just as text can be rhetorical because of the way it is structured and presented, so visuals can use color and placement to influence an audience.[36] Kress and van Leeuwen's approach is described by Alan Gross and Joseph Harmon as being "grounded in the philosophy of science and synthesized from research in cognitive psychology, semiotics, and argument and narrative theory."[37] Gross and Harmon also note that science utilizes visuals to "establish scientific truth," and that establishing this truth can rely critically on a concepts' visual depiction.[38] In fact, they describe scientific arguments as being "fortified by images."[39]

Hariman and Lucaites[40] charge that visuals aren't just about clarification or even interest. Visuals carry with them the power to evoke emotional response. The authors use patriotic symbols as an example.[41] Although Hariman and Lucaites work with iconic still photos,[42] not moving pictures, they successfully argue that the use of visuals are analogous to creating an emotive appeal. When images move, this appeal becomes even more immediate; they become "dynamically variable."[43] After all, spoken language has always been visual, too. Facial expressions, hand gestures, eyerolls: interpretation of these acts of visual communication leads to interpretation of the words.[44] Hariman and Lucaites also use the term "imagetext" to describe the visual as artifact and as purveyor of meaning.[45] This term indicates that the visual and the verbal (or textual) are inextricably intertwined. Visual representations are often steeped in *pathos*, evoking some emotional reaction from the audience. This "aesthetic engagement"[46] of the audience is essential to the rhetorical effectiveness of imagetexts. They have to be attractive and interesting in order to capture the audience's attention. In a way, visuals can also become performative guides that direct audience response.

Visual communication is highly effective because the human brain is designed to take it in and interpret it. Zender and Crutcher point out that much of the "cerebral cortex is devoted to processing visual information."[47] It can stimulate the cerebral cortex in a way that makes the information easier to understand and therefore more engaging or interesting for an audience. According to Francis Bacon, rhetors who want to put abstract information into a more "concrete" or understandable form need to translate or adapt the information into a visual form.[48] The visual can better prepare audiences to internalize or comprehend the information presented. Thomas Lessl argues that popularized science in particular, if it is to be well-received, involves a process of preparing people to receive it, not simply "translating" the data and putting it in front of them.[49] For larger audiences, like those that may be attracted by television programs, visual illustrations (combined with a display of expertise) provide a greater persuasive pull and serve to contextualize information for the audience.[50] The information is more interesting and palatable and therefore more likely to attract an audience's attention.

ĒTHOS, FLOW, AND TELEVISION

People perceive the world through individualized viewpoints that have been built through experience and education. These viewpoints determine an audience's perception of "truth," creating a great challenge to rhetors. This process of individuated perception would have been referred to as a "frameworks"[51] by Michael Polanyi or "terministic screens"[52] by Kenneth Burke. These frameworks shape the way audiences perceive information and what information they focus on, resulting in what Polanyi calls subsidiary awareness and focal awareness.[53]

Polanyi uses the example of driving a nail with a hammer to clarify the difference between the two. The act of hammering the nail focuses our attention on the handle in our hand (and the force running through it), making it foremost in our consciousness. We know the room around us is still there, but it is not our focal point and so it effectively disappears into the unconscious. Our attention is focused on only the hammer; its force and position; and our hand. Our perception of our surroundings has narrowed as we have selected only particular elements (or a single activity) on which to concentrate. The hammer is in our focal awareness; the room fades into the background—but it is still there. Mihaly Csikszentmihalyi draws parallels between this phenomenon and R. G. Collingwood's assertion that when attention is paid in one area, then it is denied in another.[54] Burke's theories about reflection, selection, and deflection of reality similarly echo this idea.

Fully understanding a single audience member's particular framework would be highly improbable; addressing the individualized frameworks for multiple people becomes exponentially more improbable. Bormann even acknowledges this difficulty: "Granted that every individual has unique experiences, the question is how do they manage to communicate with one another, cooperate in social endeavors, and come to share symbols and interpretations of their existence?"[55] This is one of the challenges for rhetors, who must get and keep the attention of an audience in order to persuade them.

Attention can't be demanded;[56] it can only be present if the audience willingly gives it. Csikszentmihalyi claims that awareness and even attention itself is "an adaptive tool of persons interacting in their environment."[57]

While interaction may seem to be a word denoting physical activity of some kind, interaction can also involve mental or emotional stimulation with little to no external movement. This stimulation is well-sprung from outside the audience and is purposely manipulated or evoked by the rhetor through building a forum (*ēthos*). The sense of shared embodiment (Sullivan's consubstantiality) that a forum creates is common with images. According to Hariman and Lucaites, "All media need virtual embodiment and interaction if they are to create a sense of shared experience and secure the continued attention of the audience."[58]

Providing pleasant experiences or attractive things to look at helps to create unbroken, focused attention, in other words, flow. The experience itself is considered the rewarding or motivating factor;[59] there are no other reasons for the pursuit other than the activity for its own sake. This caveat to creating flow may be why some people find reading novels for literature classes particularly difficult. If a student likes to read and is practiced at it, then obtaining flow is possible. Even though a book may have been assigned for a class, if the student finds the book interesting, then the motivation can shift and come to extend from the student's own pleasure in reading it. The many hours spent reading can pass quickly then, an attribute of deep flow. However, if the student doesn't like to read, doesn't find the book interesting, and/or is only motivated externally—not by the enjoyment of reading itself, but only by the need to complete the assignment—then achieving flow will be impossible. Maintaining concentration and interest will be extremely difficult and the hours invested in the project will drag.

Watching television is done usually as a hobby, an indulgence that is pursued for its own sake. It has the advantage of attracting people who will already be predisposed to achieving a state of deep enjoyment because watching television is something they like to do. Television is able to quickly change *projected* physicalities (locations) and to display symbols and rhetorical visions. The forum of television creates the same opportunities for flow no matter the topic at hand—as long as it attracts and holds the audience's interest. Watching the recording of a loved one's memorial service may be nearly as emotional—evoke the same emotions at nearly the same level of intensity—as actually attending the service may. Televised royal weddings and funerals, processionals, and memorial services for the famous (President Kennedy, Princess Diana, Michael Jackson) draw huge audiences because people want to participate virtually in the rituals. The televised events may even provide the audience with emotional release. Television invites the audience to experience consubstantiality with others who share the symbolic vision of the people involved or of the ceremony witnessed. Television brings immediacy to these events even though they are "experienced" from a distance. Flow makes the other externals in the proximity of the audience fade away, and it directs attention only to the event being shown. In this way, television is able to trigger psychological or emotional connections quite easily—as long as flow is not interrupted.

Interruptions and Flow

Interruptions can be very detrimental to achieving or maintaining a state of flow. Highly focused audiences experience a particular kind of connection to a program or forum. When an external event or action breaks into that focus, then the connection the audience felt is (at least) temporarily lost. Frustration can easily arise. The audience may become distracted in the interim and switch their focal awareness to something else. Watching an internet video and having it stop to buffer could break or dissolve the audience's interest in the video itself. The length and frequency of the buffering will affect the audience's reaction, as will individuals' levels of patience.

Commercial breaks can also interrupt flow. These breaks are part of commercial television as a genre, so tolerance for them may be greater than for video buffering. However, the popularity of DVRs (and the ability to fast forward through these interruptions) indicates that the breaks are not welcome and detract from enjoyment of the program. "Binging" on television series when available on streaming services is becoming ever more popular. Streaming (when unwelcome buffering does not become an issue) helps to avoid breaking the sense of flow with the storyline, since there are no accompanying commercials and no delays between episodes or seasons.

Cosmos was fortunate enough to appear on public television without commercial breaks. The forum it built, the *ēthos*, had the opportunity to create a sense of flow throughout entire episodes. Even at the time, journalists recognized the inherent advantages to *Cosmos* appearing commercial-less on PBS.[60] Commercials threaten to introduce any subject or tone in the middle of an attempt to build *ēthos*. Even a very serious, highly emotional frame can be interrupted for the shilling of personal care products, holiday sales, or proffered services—breaking flow. Although this book credits television as having a great ability to create a forum and establish *ēthos*, it should be noted that Cskiszentmihalyi describes TV as a medium that may challenge an audience's ability to achieve flow.[61] I assert that it is commercials and other programmatic interruptions that break into the forum and cause this interruption of flow, not the medium itself.

Images and Virtual "Reality"

Television provides excellent opportunities for *ēthos*-building because it is able to portray a virtual space that "feels" real to viewers who willingly suspend their disbelief. Through imagery and sound, music and pictures, speed and repetition, people can become quickly "habituated" to the new environment created by the medium. Television can build a forum from which audiences will view the information that a speaker presents. Because of that, they are more open to persuasion since the program's framing of the subject matter becomes the point of view from which they receive the information.

Images are an easy way to communicate meaning.[62] Since they are easily and willingly taken in, television (as a highly visual medium) enjoys an advantage for persuasion. Kress and van Leeuwen point out that appearance is the "dominant criterion" for determining what is real and what is not.[63] They argue that we determine the veracity of a visual depiction from our previous experiences; whether or not the image is deemed to be "believable" will be based on previous similar experiences.[64] Of course, when the only frame of reference is other imaginary depictions (like movies set in space), that understanding becomes skewed and audiences may determine a visual is "true" because they've seen one like it before in similar fictional or imagined settings.

In order to capture the most "belief" and cooperation from the audience, the generous use of imagery is helpful. Anthony Blair argues that the images shown even in nightly news programs are a selection of reality, manipulated to fit within a newscast and broken by voiceovers.[65] Yet, because of "the sense of realism that the visual conveys" audiences feel as though they've been given "direct access" to whatever is being shown.[66] Print is unable to convey this same feeling of presence and access—and the feeling or impression of having been there is key[67] to creating this type of shared or virtual forum. The images depicted on news programs are hardly the same as "being there," so although the filter is different between static print and "moving" film, a filter still exists. However, audience awareness of the filter's presence is greatly lessened in moving visual media.

In other words, images can lie. A camera may capture an event as it happens, but the people in control of the camera and the resultant images can alter them.[68] When the photos have been altered, then the "reality" portrayed becomes tinged with fictionalization. Even though people now often assume that images might have been "altered or fabricated,"[69] when audiences are in a state of deep flow, they are unlikely to critically question the images appearing on their screen. They are more likely to keep moving along with the story that interests them rather than breaking down individual shots or sequences for realism.

Through lulling audiences into a state of flow, a rhetor is creating a forum for persuasion and triggering a pre-existing tendency to believe what is seen. According to Umberto Eco, when messages seem to serve a "practical" purpose, the form in which the message itself is delivered can become secondary or even go unnoticed.[70] Audiences therefore may not notice when a picture is actually a computer graphic (an artistic rendering) of what a nebula *might* look like. They will take in the information and not think about the form in which it is being delivered when the images presented look real.[71] However, if the image does not look like a reflection—if it appears to be an artistic creation or interpretation of some sort—the audience may become more critically aware of the intervention of the rhetor.

The Mirror Effect

Another reaction images can elicit is called the mirror effect. Certain pictures can affect people on a psychological and even physiological level. For example, if one person sees another biting into a sour pickle (a lemon, a grapefruit, etc.), the observer's salivary glands may be activated. Neurons also can fire upon simply *seeing* an object, or even upon *seeing someone else* act upon it.[72] This kind of reaction indicates that the brain can sometimes "mistake" the observed for the experienced. Mirror neurons can make people feel victorious when a favorite athlete in the Olympics earns a gold medal. They can make people's throats start to hurt after attending a lecture given by someone who is clearly in the grips of laryngitis. Their neurons are focused on a particular action and reaction, and as a result, their own emotions and sometimes their bodies react as though they are experiencing the same event.

The mirror effect helps rhetors to create a consubstantial (shared) experience through the forum they build. Undoubtedly, it also provides additional veracity to their arguments when they are able to trigger not just intellectual reactions in their audience, but emotional and physical ones as well. Mirroring happens internally and unintentionally, an automatic reflex, like breathing. However, it can be manipulated by a rhetor who understands what triggers this reaction.

ĒTHOS, SETTING THEMES, AND FANTASY

Ernest Bormann's original 1972 article on fantasy-theme analysis explains how to identify fantasy setting themes. These settings are "given supernatural sanction," become "sacred ground" or are made "profane."[73] *Cosmos'* main setting themes throughout the program involve places that have been made sacred. Anywhere that life inhabits, anywhere that Sagan dwells, becomes sacred. Destruction or damage is an affront not only to the location itself but to the universe as a whole and to Sagan as a deified leader.[74] Specific settings that are framed as sacred include space (its exploration and potential), Sagan's ship, and the Earth.

Sacred Space

Cosmos frames settings in space as sacred, but these forums also reference what has already become sacred to the audience. They are used to persuade the audience to watch, providing *Cosmos* with the opportunity to create a setting theme and form its own symbolic cues with the audience. These settings connect the *Cosmos* journey to entertainment like *Star Wars* and *2001: A Space Odyssey*, as well as with the rhetorical vision of the program as a whole. Purposeful visual echoes of these popular movies made the program especially enticing in 1980 since it was something that audiences could watch in their own homes. This sort of connection between genres is common in fantasy-theme analysis. Bormann, Cragan, and Shields establish that "rhetoricians often use an archetypal fantasy as a way to fit the . . . unfolding of experience comfortably into the community's social reality."[75] The fantasy themes present an "organized and artistic" alternative to the way audiences experience the present.[76] They become an escape, a fantasy-based rhetorical vision. Audiences do not need to go to a theater nor pay for a ticket to join on Sagan's space journey. They are able to participate from within the comfortable environs of their own home. They are able to enter a state of flow as they experience much that is "alien" to their ordinary existence.

Space also provides a backdrop for visions of the sacred future. Sagan frames progress to the future as the only way the human race can survive. If the human race stays static, does not evolve past what he refers to as the "reptilian brain," then the species is doomed. It is only by making it through the present that humans can transcend their failings and meet a sacred destiny as explorers of the universe. As the only intelligent life that is known to exist in the universe, humans owe a sacred duty to the cosmos itself to live and to evolve. These visions are frequently drenched with images of travel to other planets.

This setting theme also includes looking at the Earth as "outsiders." In the fifth episode, "Blues for a Red Planet," Sagan asks the audience to view Earth from the perspective of aliens. He directs,

> Now, imagine yourself a visitor from some other and quite alien planet. You approach the earth with no preconceptions. Is the place inhabited? At what point can you decide? When we look at the whole Earth, there are no signs of life. We must examine it more closely.

While he is speaking, a picture of the Earth, round and blue, is displayed outside the ship's window. As the picture shifts to show the Earth from "a few kilometers" away, Sagan declares that Earth still appears to be "utterly lifeless." It is not until he has brought the ship in closer still, to the point where details "as small as one-hundred meters across" appear, that the planet begins to show the telltale patterns of intentional design. The point is clear: life is hard to see from too far away. In order to determine visually whether intelligent life exists on other planets, the audience will have to get much closer to those planets. It's a conclusion that can be reached independently by audience members because of the forum (*ēthos*) Sagan uses. He has shifted the vision of the audience to look at information in a particular way.

This scene changes at about minute 26 to an examination of the surface of Mars. The pictures of this planet's surface appear to be actual photographs, most likely taken by NASA, but around minute 28, the images are again displayed as though through the front window of Sagan's Spaceship of the Imagination. The images are tinted reddish and they depict the surface from a much closer range. These computer-created animations demonstrate what scientists believe the Martian surface might look like if its image were captured during such a "flyover."

The ship "glides" across the surface and close up to the Mariner Valley. At 29:30, Sagan is standing in front of the ship's window and it appears that the ship is traveling swiftly through the valley. The images are reminiscent of the final battle scene in *Star Wars* in which Luke Skywalker flies through the Death Star's trench. However, in that scene, Luke is on a mission of destruction, but Sagan is on a fantasy-filled mission of discovery. The red glow from the planet is cast over everything in Sagan's ship. After a minute of this virtual experience, Sagan states, "To skim over the sand dunes of Mars is, as yet, only a dream." Because Sagan notes that "robot emissaries" have already been sent to Mars, the impression left is that these graphics are likely realistic depictions of the planet.

After this adventure, Sagan discusses the very real Viking mission to Mars. He shares a map and tells where the craft landed, then the scene shifts again at about minute 32 to show what appears to be a volcanic eruption. Sagan says, "So, after a voyage of 100 million kilometers, on July 20, 1976, Viking 1 landed right on target in the Chryse Plain." The image shown looks like a view from the Viking as it lands, although this would be impossible. Stones are blown out of the way from the exhaust as the probe lands. The image changes to show the outside of the vehicle as it "awakens," and Sagan tells what the Viking did upon reaching the planet's surface.

The program combines narrative about the real, documented event alongside *constructed* imagery, blurring the line between what really happened and what *Cosmos* imagines it looked like. The artistically rendered images are all bathed in the reddish hue stereotypically associated with Mars. At 33:40, the program shares an actual black and white image taken by Viking of its own "foot," the first image ever sent from the planet's surface. However, the very next picture is from a set recreation of the landing. It is colored in the same way as the real pictures of Mars that follow. An audience experiencing deep flow is unlikely to discern this difference and is likely to have taken all the photos in as "fact." The intermixing of photograph with artistic representation (or re-enactment) blurs the line between the two.[77] The forum created by the program presents a singular way in which to take in and consider the information. Since one image follows another and none appear less "real" than the others, the audience does not have reason to doubt or to assume that some of what they've seen is fact and some is supposition. The forum creates trust and focus, removing doubt and encouraging belief. Mars becomes a fantasy setting that represents "sacred ground." Its exploration is one step into the greater cosmos, supporting the program's primary action theme of space exploration.

Sagan's Sacred Ship

The Spaceship of the Imagination serves as a fantasy-setting theme as it habituates the audience to a particular view of space and space exploration. The shining, dandelion-shaped exterior appears in all the episodes during the opening credits. The interior of the ship appears in at least nine of the thirteen episodes. When it is present, Sagan discusses topics like "the stuff of life" or the birth of stars;[78] compares earth's environment and possible destruction with that of other planets and civilizations; invites viewers to examine earth from the outside, as aliens;[79] explores the potential for future space travel and the boundaries of the universe;[80] and makes connections between the modern audience and the astronomers of our ancient past.[81] By maintaining a similar forum for examining a variety of arguments, *Cosmos* removes some of the alien or uncomfortable elements, which may bring the audience out of the program—or break flow—and allows the audience to examine new or controversial ideas from a comfortable and familiar location.

The ship in this way provides not just a frame for argumentation, but a lens through which the audience will perceive the information conveyed. The ship as a setting provides comfort by being a touchstone to which the show returns repeatedly and reliably. Even though each episode has a different focus and may take the audience to any number of foreign or "alien" places, this place remains the same, sacred, inviolable. The ship becomes a home-base of safety within exploration. Each time it is shown, only Sagan is there.[82] The audience is never jarred by the sudden appearance of another captain or a new navigator—or even another passenger. The person in whom the audience has placed trust continues to earn that confidence by fulfilling the role established.

If space is sacred, the ship becomes the audience's sacred place of worship. The white cathedral-like interior is a fitting environment for the omniscient being Sagan comes to represent in the show.[83] The ceiling of the ship curves upward, creating an unseen peak at the top. The curved "beams" in the walls form opaque, stylized "windows" that are filled with white rather than stained glass. Colors in Sagan's forum are reserved for life; white, on the other hand, represents the future. It carries with it symbolism from other movies. For example, in *2001: A Space Odyssey*, ship interiors were predominately white. Rather than signifying a future devoid of color, the use of white symbolizes a clean, idealized future. The future is neatly packaged, hermetically sealed, and germ-free. Dust and bacteria are a concern for less-evolved environs. In *2001*, technology and general human advancement have managed to create an environment as sterile as a laboratory. In *Cosmos*, this sterility is given more spiritual substance with its church-like architecture or design. White becomes a color not only of cleanliness; it comes to represent purity. As a figurative "heaven," this ship is the reward for correct behavior; it has been earned through following the prescripts of our divine leader.[84] In addition, whiteness of the interior indicates a certain purity. Nothing is "hidden" in busy patterns or overpowering colors. The ship is simple but not barren. The space—this forum—is neutral. It won't "clash" with color preferences or become quickly dated.

The ship's dim lighting is reminiscent of a confessional. Repentant sinners enter the booth that provides some illusion of "disguise," making them more comfortable when admitting their transgressions. This lighting contrasts with the bright sunlight that comes between the leaves in Sagan's "tree of life."[85] At the same time, the whiteness of Sagan's surroundings contrast with the deep greens of nature, the blue of the water, and the brown of the sand. Even though the ship represents a haven for the audience as aspiring adventurers, it contrasts with the living hues of nature. Color is reserved for the Earth, other planets, and space. Just as the ship's stylized "windows" are opaque and white rather than translucent stained glass, so the window at the front of the ship lets in all the light and color that the cosmic journey has to offer. Through this vehicle, the audience can learn about nature and value both life and intelligence. Sagan's control panel serves as an "altar," the gateway to discovery and human salvation. The dimmed light in turn also allows the audience the "anonymity" to question what otherwise might be blasphemous: what is the origin of life? Who is in control of life, the universe, time, and the future?

The setting of the ship, its layout, color, and repeated visits, build a complex, consistent, web-like forum (or frame) throughout the entire series. In fact, at the end of the final episode, the very last thing the audience sees before the credits roll is the empty chair, at the front of the ship. The stars through the front window indicate continued forward movement, but no one is at the helm. The implication here, through this frame, is that the audience is now capable of steering the ship themselves. They no longer require someone to direct them through the journey. They are now well informed and can be trusted to make decisions on their own. Some sense of loss may be communicated through the empty chair; the audience's guide is now gone. However, the setting has communicated the program's rhetorical vision in its last shot: evolution beyond the status quo is crucial and even looming. The audience is given the privilege of free will and control of its own destiny as a species.

The Sacred Earth

Earth represents the sacred setting for life—potentially all life—in the universe. It is also the home of color; the most dynamic contrasts are presented here. When the *Cosmos* journey visits earth, the scenes are filled with color and sometimes noise. This noise can be human-generated, as by the street cars in "The Backbone of Night,"[86] or by nature, such as the waves against the side of the ship in "The Persistence of Memory."[87] The earth setting is depicted as being endangered by people. Sagan emphasizes that the only known intelligent life in the universe exists on Earth, and the planet is being destroyed by that "intelligence." He frames the destruction as entirely avoidable and the audience may perceive from his lessons that they are destroying their planet, their "real-life spaceship," only to their utter doom.

Certain places on earth are given additional sacred emphasis in *Cosmos*. One of these places is the ocean. It is the forum from which the program addresses the origins of life and evolution. The seaside setting in Monterey, California, with its picturesque cliffs, whipping wind, and crashing waves, bookends the program. The audience first meets Sagan in this place, and it is here where the audience leaves him at the end of the program. While the final image of *Cosmos* is of the interior of his ship, Sagan is not present. The last place where Sagan actually appears is near the water—the place he uses as a metaphor for life, exploration and evolution. From these sacred shores, he discusses the evolution of life on Earth and the exploration of this planet by seafarers of centuries past.

The ocean also lends beauty and an element of awe, perhaps even reverence, for the beauty of sacred Earth. The setting creates an anchor to home and the familiar, and from here Sagan launches an adventure based in the realm of "imagination." This frame of the familiar may lessen the trepidation and pause that the audience may have otherwise had regarding extending themselves beyond their comfort zones of belief. Sagan sometimes uses this forum to depict the unfamiliar and the (potentially) objectionable; it directs the vision of the audience and provides a "tone" for their reception of the information.

Earth and the ocean shore are sacred settings on their own, but they are connected directly to the sacredness of space as well. Sagan uses exploration of Earth's oceans as a clear metaphor for space exploration. In "Traveller's

Tales," he discusses the benefits of Holland's investment in exploratory seafaring missions: "Never before or since has Holland boasted such a galaxy of scientists, mathematicians, philosophers, and artists." This investment led to great societal benefits and advances. Later he reveals that these ships were called "flying boats," and then he declares the "Voyageur spacecraft are their descendants." By framing the information from this familiar shore, which serves as the edge of "the cosmic ocean,"[88] he frames the investment in space exploration as equivalent in importance to these sea voyages. He even notes that it is now[89] cheaper to invest in space initiatives (adjusted for inflation) than it was to invest in these sea voyages in the seventeenth and eighteenth centuries. He also uses the forum to address the audience's potential economic concerns of investing money during a bad economy on something that may seem like a nonnecessity. Sagan frames these voyages as the way out of difficult times, as a way to usher in a more affluent age.

Final Thoughts: Setting Themes and Popular Entertainment

Sagan and the creators of *Cosmos* understood the great challenge in front of them: they needed to keep the attention of a broad audience while presenting thirteen hours of science programming over a half-season of television. In order to address this challenge, *Cosmos'* settings were created purposefully to reflect the science fiction and "space adventure" entertainment of the time. Piggybacking on the immense popularity of *Star Wars* attracted an audience that was potentially predisposed to appreciating this type of forum or presentation. *Star Wars* attracted huge audiences; its adventures center on space, space travel, and alien civilizations—and these topics are what *Cosmos* offers in abundance.

In September 1980, the month when *Cosmos* premiered, only two parts of the original *Star Wars* trilogy had been released. The series' storylines and characters had captured the attention of audiences across the entertainment universe. Fans were looking back at the first two movies and looking forward to the third, which would be released in 1983. It was the birth of the *Star Wars* phenomenon, when George Lucas was idolized as an entertainment god, and any references to that franchise were likely to garner attention, no matter how the franchise name was being invoked. A *Time* cover story, which was published during *Cosmos*' first run on television, declares that the program "features special effects rivaling those in *Star Wars*: computer animation, scale models and painted backdrops as dazzling as anything ever attempted on television."[90] *Cosmos* actively references *Star Wars* in its promotional blurbs and at times echoes it through visual effects. These references capture and focus the attention of its audience.

Focused attention, which can be greatly enhanced by an intellectual or emotional pull, provides opportunity for persuasion. The most engaged state of attention is sometimes referred to as a state of deep flow, or a state of focused engagement that effectively erases outside distractions, including the passage of time.[91] Nakamura and Csikszentmihalyi point out that a state of deep flow translates to "deep enjoyment."[92] One cannot exist without the other. The experience of intense, pleasurable concentration "is described across cultures in very similar ways," although triggers for it are individualized.[93] In other words, what interests one population may not interest another, but humans all have activities or subjects that can move them to a state of focused concentration in which the rest of the world fades away.

The visual effects of television create a highly effective rhetorical forum that can elicit a state of flow. Intricate, artistic, dazzling effects are both interesting and engaging and so they are often able to capture the attention of an audience. When that attention becomes a state of deep flow, or when the audiences gives their "undivided attention"[94] as Yoos would say, then a forum for persuasion has been created. Yoos calls this sort of forum-building a "genuine ethical appeal,"[95] or what this book identifies as *ēthos*. Much like the audience provides *ethos* to the rhetor if it has been sufficiently earned, so the audience also gives the rhetor their attention if it has been piqued.

Through submitting to the forum, the audience can feel as though they are experiencing the setting or events depicted on the screen. Csikszentmihalyi credits William James with noting that interest is key in creating experience.[96] If audiences aren't interested in something—whatever that something is—then they can't *experience* it. Deep flow allows audiences to be immersed in the experience the rhetor is providing. Visual effects are intended to both attract attention and to hold it until flow is achieved. In turn, a state of deep flow provides the opportunity for building a resonant fantasy theme.

Cosmos' virtual forum presents a friendly, unthreatening view of space and of science. Visual appeal is key to making this educational program transcend the ordinary boundaries of science education and enter the realm of fantasy entertainment. *Cosmos'* fantasy settings rely on visual elements to frame space adventures and take its viewers on voyages across galaxies. Through the creation of a forum, the audience is able to join and participate in the fantasy.

NOTES

1. Ernest G. Bormann, John F. Cragan, and Donald C. Shields, "Three Decades of Developing, Grounding, and Using Symbolic Convergence Theory (SCT)," in *Communication Yearbook 25*, ed. William B. Gudykunst (Mahwah, NJ: Lawrence Erlbaum, 2001), 273.

2. Ernest G. Bormann, "Fantasy and Rhetorical Vision: Ten Years Later," *Quarterly Journal of Speech* 68 (1982): 294, doi:0.1080/00335638209383614.

3. Ernest G. Bormann, "Fantasy and Rhetorical Vision: The Rhetorical Criticism of Social Reality," *Quarterly Journal of Speech* 58, no. 4 (1972): 401, doi:10.1080/00335637209383138.

4. Adrian Malone, "Who Speaks for the Earth," *Cosmos: A Personal Voyage* (Studio City, CA: Cosmos Studios, 1980).

5. Lawrence Badash, "Nuclear Winter: Scientists in the Political Arena," *Physics in Perspective* 3, no. 1 (2001): 87.

6. Ernest G. Bormann et al., "Rhetorical Visions of Committed Voters: Fantasy Theme Analysis of a Large Sample Survey," *Critical Studies in Mass Communication* 1 (1984): 287–310, doi:10.1080/15295038409360038; Ernest G. Bormann, "Symbolic Convergence Theory: A Communication Formulation," *Journal of Communication* 35, no. 4 (December 1985): 128–38; Ernest G. Bormann, John F. Cragan, and Donald C. Shields, "In Defense of Symbolic Convergence Theory: A Look at the Theory and Its Criticisms After Two Decades," *Communication Theory* 4, no. 4 (1994): 259–94; Bormann, "Fantasy and Rhetorical Vision: The Rhetorical Criticism of Social Reality"; Bormann, "Fantasy and Rhetorical Vision: Ten Years Later."

7. George Yoos, "A Revision of the Concept of Ethical Appeal," *Philosophy and Rhetoric* 12, no. 1 (1979): 46.

8. Ibid., 57.

9. Ibid., 41.

10. S. Michael Halloran, "Aristotle's Concept of Ethos, or If Not His Somebody Else's," *Rhetoric Review* 1, no. 1 (1982): 60, http://www.jstor.org/stable/465559.

11. Ibid.

12. Dale Sullivan, "The Ethos of the Epideictic Encounter," *Philosophy and Rhetoric* 26, no. 2 (1993): 127.

13. Charles Chamberlain, "From Haunts to Character: The Meaning of Ethos and Its Relation to Ethics," *Helios* 2, no. 11 (1984): 98.

14. Food insecurity is an emotional reaction to unreliable access to food.

15. Ibid., 14.

16. The section entitled "The Rhetoric of Place and Television" discusses the ways that television mirrors the attributes of physical space, as defined by Endres and Senda-Cooke.

17. Michael J. Hyde, "Rhetorically, We Dwell," in *The Ethos of Rhetoric*, ed. Michael J. Hyde (Columbia, SC: University of South Carolina Press, 2004), xiii.

18. Ibid.

19. Ibid., xvi.

20. Ibid., xvii.

21. Ibid., xviii. "The moving of the passions . . . is a prerequisite of persuasion; truth alone is not sufficient to guide the thoughtful actions of human beings."

22. Ibid., xxi.

23. Danielle Endres and Samantha Senda-Cook, "Location Matters: The Rhetoric of Place in Protest," *The Quarterly Journal of Speech* 97, no. 3 (2011): 259.

24. Ibid.

25. Ibid. Emphasis in the original.

26. Ibid. Emphasis in the original.
27. Ibid. Emphasis in the original.
28. Marcel Chotkowski La Follette, "Multivariate Perceptions: Reflections on the 147th Annual Meeting of the American Association for the Advancement of Science," *Science* 6, no. 35 (1981): 33.
29. Ibid.
30. *ET: The Extra Terrestrial* went through "ideological retooling" when Steven Spielberg replaced the guns in agent's hands with walkie-talkies.
31. Sullivan, "The Ethos of the Epideictic Encounter."
32. "*Cosmos* for Another Generation" explores *Cosmos* 2014 in more depth.
33. Chaim Perelman and Lucie Olbrechts-Tyteca, *The New Rhetoric: A Treatise on Argumentation*, ed. John Wilkinson and Purcell Weaver (London: University of Notre Dame Press, 1969), 358.
34. Gunther Kress and Theo van Leeuwen, *Reading Images: The Grammar of Visual Design* (New York: Routledge, 1996), 1.
35. Ibid., 7. The quote reads: "Sign-makers use the forms they consider apt for the expression of their meaning, in any medium in which they can make signs."
36. Ibid., 2.
37. Alan G. Gross and Joseph E. Harmon, *Science from Sight to Insight* (Chicago: University of Chicago Press, 2014), 10.
38. Ibid., 1.
39. Ibid., 6.
40. Robert Hariman and John Louis Lucaites, *No Caption Needed* (Chicago: University of Chicago Press, 2007).
41. Ibid., 1.
42. Ibid., 7. Hariman and Lucaites report: "The photos we have selected not only are widely recognized but also are thought to have had distinctive influence on public opinion."
43. Kress and van Leeuwen, *Reading Images: The Grammar of Visual Design*, 242.
44. Ibid., 39.
45. Hariman and Lucaites, *No Caption Needed*, 24.
46. Ibid., 30.
47. Mike Zender and Keith A Crutcher, "Visible Language for the Expression of Scientific Concepts," *Visible Language* 41, no. 1 (2007): 26.
48. Karl R Wallace, *Francis Bacon and the Nature of Man* (Urbana: University of Illinois Press, 1967), 80.
49. Thomas M Lessl, "The Priestly Voice," *Quarterly Journal of Speech* 75, no. 2 (1989): 195.
50. Mark Meister, "Meteorology and the Rhetoric of Nature's Display," *Quarterly Journal of Speech* 87, no. 4 (2001): 415–28.
51. Michael Polanyi, *Personal Knowledge: Towards a Post-Critical Philosophy*, Harper Tor (New York: Harper & Row, 1964).
52. Kenneth Burke, *Language as Symbolic Action* (Berkeley, CA: University of California Press, 1966).

53. Polanyi, *Personal Knowledge: Towards a Post-Critical Philosophy*, 55.

54. Mihaly Csikszentmihalyi, "Attention and the Holistic Approach to Behavior," in *Flow and the Foundations of Positive Psychology: The Collected Works of Mihaly Csikszentmihalyi* (Claremont, CA: Springer, 2014), 3.

55. Bormann, "Fantasy and Rhetorical Vision: Ten Years Later," 295. Footnote.

56. Csikszentmihalyi, "Attention and the Holistic Approach to Behavior," 9.

57. Ibid., 6.

58. Hariman and Lucaites, *No Caption Needed*, 41.

59. Jeanne Nakamura and Mihaly Csikszentmihalyi, "The Concept of Flow," in *Flow and the Foundations of Positive Psychology: The Collected Works of Mihaly Csikszentmihalyi*, ed. Springer (Claremont, CA, 2014), 240.

60. Judy Klein, "The Medium Gets a Message," *Science News* 115, no. 22 (1979): 361.

61. Csikszentmihalyi, "Attention and the Holistic Approach to Behavior," 10.

62. Kress and van Leeuwen, *Reading Images: The Grammar of Visual Design*, 11. Quote reads: "Communication requires that participants make their messages maximally understandable in a particular context. They therefore choose forms of expression which they believe to be maximally transparent to other participants."

63. Ibid., 163.

64. Ibid.

65. J. Anthony Blair, "The Rhetoric of Visual Arguments," in *Defining Visual Rhetorics*, ed. Charles A Hill and Marguerite Helmers (Mahwah, NJ: Lawrence Erlbaum, 2004), 51.

66. Ibid.

67. Ibid.

68. Kress and van Leeuwen, *Reading Images: The Grammar of Visual Design*, 159.

69. Hariman and Lucaites, *No Caption Needed*, 23.

70. Umberto Eco, *Experiences in Translation*, ed. Alastair McEwen (Toronto: University of Toronto Press, 2001).

71. Ibid., 93.

72. "Mirror Neurons," *Principles of Neural Science*, 2013, 422.

73. Bormann, "Fantasy and Rhetorical Vision: The Rhetorical Criticism of Social Reality," 401.

74. Sagan as divine is explored in chapter 3, *Ethos* and the Popular Science Persona

75. Bormann, Cragan, and Shields, "In Defense of Symbolic Convergence Theory: A Look at the Theory and Its Criticisms After Two Decades," 281.

76. Ibid.

77. "*Ethos* and the Popular Science Persona" discusses this idea in more depth.

78. Adrian Malone, "Shores of the Cosmic Ocean," *Cosmos: A Personal Voyage* (Studio City, CA: Cosmos Studios, 1980); Adrian Malone, "One Voice in the Cosmic Fugue," *Cosmos: A Personal Voyage* (Studio City, CA: Cosmos Studios, 1980); Adrian Malone, "The Lives of the Stars," *Cosmos: A Personal Voyage* (Studio City, CA: Cosmos Studios, 1980).

79. Adrian Malone, "Blues for a Red Planet," *Cosmos: A Personal Voyage* (Studio City, CA: Cosmos Studios, 1980).

80. Adrian Malone, "Traveller's Tales," *Cosmos: A Personal Voyage* (Studio City, CA: Cosmos Studios, 1980); Adrian Malone, "The Edge of Forever," *Cosmos: A Personal Voyage* (Studio City, CA: Cosmos Studios, 1980).

81. Malone, "Who Speaks for the Earth."

82. Sagan is present in every broad interior shot of the ship except the final one.

83. This idea is explored in chapter 3, "*Ethos* and the Popular Science Persona."

84. Sagan's role in the program is discussed in chapter 3, "*Ethos* and the Popular Science Persona."

85. Malone, "The Edge of Forever."

86. Adrian Malone, "The Backbone of Night," *Cosmos: A Personal Voyage* (Studio City, CA: Cosmos Studios, 1980).

87. Adrian Malone, "The Persistence of Memory," *Cosmos: A Personal Voyage* (Studio City, CA: Cosmos Studios, 1980).

88. Malone, "Shores of the Cosmic Ocean."

89. 1980

90. Frederick Golden, "The Cosmic Explorer," *Time*, October 1980, 63.

91. Nakamura and Csikszentmihalyi, "The Concept of Flow," 240.

92. Ibid., 241.

93. Mihaly Csikszentmihalyi, "Flow with Soul," in *Flow and the Foundations of Positive Psychology: The Collected Works of Mihaly Csikszentmihalyi* (Claremont, CA: Springer, 2014), 273.

94. Yoos, "A Revision of the Concept of Ethical Appeal," 57–58.

95. Ibid., 57. Yoos notes that "genuine ethical appeal[s]" [*ēthos*] are those that will capture the "undivided attention" of the audience.

96. Csikszentmihalyi, "Attention and the Holistic Approach to Behavior," 3.

Chapter 5

Cosmos and the Creation of a Popular Science *Mythos*

In ancient Greece, myths formed a "sacred narrative explaining how the world and man came to be in their present form."[1] These myths served to document the values of Greek society; they were not intended to be seen as "fiction," created for entertainment purposes only. They were intricately interwoven with contemporary understanding of society. In many ways, they were as instructive as they were entertaining. Greek mythology focuses on religious imaginings of gods and their complex relationships with each other and with humans in order to explain how the world and humanity were created. These tales become in some ways sacred, evoking a "mixture of fear and attachment . . . respect, desire, and terror."[2] They are well-known, canonical, almost universally revered—not just as "stories" but as vital artifacts of civilization and humanity itself.

In order to entice people to listen, these stories had to be interesting enough to attract attention and they had to be clear enough so their lessons would be understood. Common elements that predictably resonated with an audience formed archetypes. These archetypes contained recognizable patterns of action, location, characters, or plot themes. These recurring themes became shortcuts to establishing a narrative rhythm that is familiar and easy to follow. Archetypes and myths build a recognizable scaffolding on which arguments can be discretely built. The familiarity of the structure provides comfort for the audience who may be challenged already by the arguments (or directives) embedded within the story. The content becomes (in some ways) anticipated and audience reflections on it predictable.[3] Formulaic construction, honed over centuries, can help an audience easily follow and understand plotlines or character motivations even in modern cinema and television.

Mythic tales and their archetypical structure are still relevant today. Sagan himself connected *Cosmos*, and even science itself, to myth. In a 1980 interview with *Rolling Stone*, he notes that the questions posed by *Cosmos* "are the subject of folklore, myth, superstition and religion in every human culture."[4] By selecting universal questions, asked since the dawn of humanity, Sagan directly ties the *Cosmos* journey with the ancient yearnings of the human soul—not just the contemporary human intellect. By tackling these questions, he helps the audience to wonder about, put visuals to, and contextualize concepts that are normally connected with the spiritual. *Cosmos'* mythic narrative is an origin story, meant to explain where we come from and who we are. It is reminiscent of a Homeric epic, which Paolo Vivante describes as being an elaborately intertwining genesis tale that connects people and animals to all of creation and traces their beginning.[5] Its dramatic structure builds to reinforce the program's "divine revelation."[6] This revelation is meant to resonate with audiences and impact their attitudes and even actions beyond the program.

The essence of *mythos* is a "feeling of connectedness that is characteristic of aesthetic experience," an experience that explains our world and how we fit within it.[7] This definition illuminates how fantasy themes of human evolution create *Cosmos'* resonant *mythos*. *Cosmos* tells the story of humanity and the science fiction-like narrative scaffolds the science content to make it appealing to viewers. *Cosmos'* fantasy themes echo ancient Greek myths because, like mythic tales, the program intends to educate people about creation and the origins of life. It also wants to help guide their future behavior. The message's resonance, saturated with *mythos*, is a result of the dramatically structured narrative that has been built through addressing *kairos*, *ethos*, and *ēthos*.

CREATION, EVOLUTION, MYTH, AND *COSMOS*

In modern times, the word "myth" is associated with something that is known to be untrue but that (at one time) was held as a common belief. Rather than indicting or rejecting these natural instincts to mythologize in the pursuit of answers, Sagan says that the makers of *Cosmos* "respect them as part of the same impulse that has led to the modern scientific kind of myth."[8] He declares, "I don't think you should be surprised that a religious idea turns out to have some scientific support."[9] Sagan is melding the identity of science with religion by defining "myth" as an explanation, drawn from "the best information . . . available" about "the origin of something."[10] Sagan then claims that science, based on this definition, is also myth. The myths created by science, he says, are almost "guaranteed" to eventually be replaced with something else, that the things we can't even conceive of now will be considered fact in the future.[11] He essentially frames religious or mythic explanations of our origin and creation as the first rungs on the scientific ladder.

These ideas have been replaced with modern science, he postulates, and in the same way, the ideas put forth by modern science will eventually be replaced by other things as well.

Given Sagan's position in this interview, it's fair to say that *Cosmos* was set up from the beginning to connect to the mythic through dramatization of science. To illustrate this point, the marketing boilerplate used to promote *Cosmos* can be compared to that of another popular science program, which premiered the same week as *Cosmos*. This program, *The Body in Question*, boasts a talented and intelligent host named Jonathan Miller, a "noted British physician, writer, actor, lecturer and theatre director."[12] While the accomplishments (at that time) of both Sagan and Miller seem equally impressive, a key difference between the two programs is that Miller promises education rather than exploration. For example, the first episode of *The Body in Question* is called "Naming of Parts." According to a promotional blurb in *Science,* the program "postulates that our ignorance of basic physiology leads to unusual attitudes toward our own insides. Using a variety of media, Miller explains what happens when we fall ill."[13] In contrast, on the same page in the same issue, *Science* describes *Cosmos* in this way: "Sagan employs a 'spaceship of the imagination' to take viewers on a simulated, scientifically accurate journey from the edge of the known universe 8 billion light years away."[14] Even though both programs seek to educate the public about science, their promotional blurbs indicate that their approach is very different. *Cosmos* is saturated with *mythos*—a dramatic adventure, perhaps like that of Odysseus. *The Body in Question* promises to look at "our own insides . . . when we fall ill"—promising a look at mucus formation and perhaps stomach viruses. The connection is with physicality, not with epic adventure.

In addition, while the *Cosmos* blurb emphasizes journey and discovery through simple yet poetic language, the blurb for Miller's program references the "ignorance" of its audience and it uses academic terminology. While *Cosmos* is described like a program focused on entertainment, *The Body in Question* is described like a program focused on lecture. While *Cosmos*'s description evokes images of space that are reminiscent of the contemporary entertainment juggernaut *Star Wars* and its sequel *The Empire Strikes Back*, *The Body in Question* makes no such references, either through use of language or through subject matter. Illness is its focus, not a story of excitement and exploration. Only one of these two programs was destined to capture the imagination of millions of viewers and linger in memory and popular culture for decades.

The story of humanity's origin and its evolution plays a key role in *Cosmos'* formation of *mythos*. Beginning from the first episode, Sagan proclaims that "we are made of star-stuff," a statement he repeats in the eighth episode, "Travels in Space and Time," as well as in the final episode, "Who

Speaks for the Earth." It is the audience's introduction to the exigence of the story and it is not allowed to fade from their consciousness through the program. Humanity's "oneness" with the universe is tied to its evolution from "star-stuff" to homo sapiens to the explorers humans may someday become. However, this narrative is not presented in a purely chronological manner leading to humanity as the pinnacle, as a narrative about evolution quite often would be. The human race is born out of the death of a star, elements that evolve into an ever-improving form of life. It attains a consciousness, which learns about the universe as it learns about itself, and eventually goes back out into the *Cosmos* as the same elements from which they sprung, part of the universe that permeates everything.

As Sagan describes in episode nine, "The Lives of the Stars," stars "sparked the origin of life." Elements are pushed out into the universe and they eventually form other things, sometimes living things. Some elements eventually became people. To emphasize human importance within the universe, Sagan proclaims in the first episode, "Shores of the Cosmic Ocean," "We are a way for the universe to know itself." Again, this statement is repeated in the eighth episode[15] and at the end of the series. Sagan distinguishes the audience as the consciousness of the *Cosmos* and therefore an integral part of all that is. This sentiment is also present in spirit when he declares that exploration of the universe is a mission of self-discovery.[16] To emphasize this point, he tells a school child, "You're part of the Milky Way."[17]

In the second episode, "One Voice in the Cosmic Fugue," Sagan illustrates how humans can guide evolution when he tells the story of Japanese crabs on whose shells appear markings very similar to that of a human face. Around these crabs had grown a myth that the faces were those of Heike warriors, and so fisherman would throw the crabs back to the sea when they found them in their nets. Sagan points out that humans directed this evolution—they chose to preserve the lives of these crabs and take the lives of others. In the same way, the narrative implies, humanity can choose its own destiny and direct its own evolution, just as the fisherman did with the crabs.

Because of its power, Sagan presents evolution as positive, even "uplifting."[18] He describes it as an opportunity to choose our actions as a species, as inhabitants of this planet and as part of the *Cosmos* that surrounds us. Of course, he points out that evolution has drawbacks as well, such as a still-present human (primitive) penchant for violence[19] that stems from our thus-far unevolved "reptilian brain."[20] He reveals in episode 4, "Heaven and Hell," that humans could very well "force the hand of nature," and destroy not just human life, but all life on Earth. He repeats the introductory phrase, "If we do not destroy ourselves"[21] to point out that humanity's still-evolving weakness could result in catastrophic destruction if it does not take control of itself. Last, Sagan also presents human travel beyond the limits of our small planet as a natural progression from seafaring voyages—an evolution of investigation and an evolution of the human species.[22]

One of the main thrusts of *Cosmos'* mythic narrative of evolution is the miracle and randomness of life: "There was a particular sequence of environmental accidents and random mutations in the hereditary material" which formed life.[23] This information, "from the point of view of a star," illustrates that the spark of life was not purposeful, nor did it come from "outside" as part of a plan. While people can control what happens with life once it is formed, the creation of life itself is arbitrary. Therefore, according to Sagan, there is no god to save humanity from a fall, whether that fall comes from a set of random accidents or from its own irresponsibility, carelessness, or selfishness in regard to planet Earth. When Sagan shows the line drawings illustrating human evolution, beginning with single-celled organisms,[24] he stops with the present human form. However, the implication throughout the series is that humanity is, as yet, not fully evolved as a species. His reference to the "reptilian brain"[25] makes that clear. Sagan shows that even the constellations change and evolve,[26] so humans can be no different. We are made of star-stuff. Our origin story, according to Sagan and *Cosmos*, starts and ends with the stars.

SCIENCE FICTION'S INFLUENCE ON *COSMOS*

While these dramatic vignettes provide short illustrations of information within the more complex narrative structure, *Cosmos* as a whole appears to reflect one of the most common themes within science fiction: an adventure or journey through space. As explored previously in this book, the first episode, "Shores of the Cosmic Ocean," takes viewers on a voyage "across the known universe," from the edge of space to the "local group" and the Milky Way. This voyage mirrors the one in Olaf Stapledon's iconic *Star Maker*. While the Stapledon's journey starts from Earth and heads outward and Sagan's trip starts from the edge of the universe and travels toward Earth, the comparison is striking. *Star Maker* is a meandering journey that has a loose plot, binding together shorter stories about the narrator's adventures through space and many "epochs" (a word Sagan also uses) of time. It has a strong message against war and the destruction it causes—but also about the near unavoidability of it.

Damien Broderick makes a connection between ancient myth and science fiction, claiming that "science fiction is . . . the myth-form of the industrial age."[27] While popular science and science fiction are separate genres, some overlap exists in popular science artifacts that hope to create a resonant *mythos* and have a lasting impact on society. *Cosmos* in particular invoked techniques of the science fiction genre to attract and retain an audience. From the beginning of his career, Sagan's science often was influenced by his interest in science fiction. He was a dreamer and his imagination was greatly influenced by the sci-fi of his youth.

According to biographer Keay Davidson, Sagan grew up reading science fiction, enthralled by the works of Edgar Rice Burroughs. Davidson even states that Sagan's serious proposals about "terraforming" Venus reflected this science fiction background; Jack Williamson originated the term, but the concept came from science fiction writer Olaf Stapledon, who "suggested electrolyzing Venus's supposed oceans to generate oxygen for its atmosphere."[28] Friends reported of Sagan, "He was presenting [his ideas about extraterrestrial life] as science but to me, this was science *fiction*."[29] So, it should be no surprise that the similarity of *Cosmos* to science fiction has not gone unnoticed. In 1981, Martin Green wrote, "Sagan is likely to arouse the same reactions in a man of letters as Wells and his followers have aroused."[30] As noted earlier in this book, Thomas M. Lessl claimed that *Cosmos* was a genre hybrid, mixing education with science fiction.[31] Elements of the story and its structure are clearly reminiscent of (and referential to) science fiction. H. G. Wells was an influence both directly and indirectly on Sagan and *Cosmos*.

Cosmos' story, structure, and visual depiction reference science fiction in order to connect with audiences. An extended dramatic depiction of H. G. Wells's *War of the Worlds* begins approximately two minutes into the fifth episode, "Blues for a Red Planet." It features avant-garde imagery, including what may be the mechanisms in a clock and an eye. It attempts to reproduce the "innocence" of an earthly civilization continuing on, unaware it is being watched by aliens who have an "envious eye" for our planet. Eventually, Sagan breaks into the narrative to briefly discuss the story itself and the famous radio depiction that anecdotally caused a panic among listeners some forty years after the original story was written.

Before moving on to talk about Burroughs's John Carter novels and the title character's adventures on Barsoom (Mars), Sagan presents information about astronomer Percival Lowell. This structure effectively "sandwiches" historical information about a *real* scientist investigating the *real* possibility of life on Mars between two science fiction stories about *fictional* life on Mars. Because the "true" science is placed between the fictional accounts, a fuzzy demarcation between popularized science and science fiction is created. Lowell thought life on Mars was real. He really thought that he saw water being directed through canals to civilizations in warmer, drier regions of the planet. The audience is not asked—nor is it motivated to—keep straight which of the three stories are science fiction and which represent authentic scientific hypothesis. The background of science fiction injects interest and drama to the account of scientific history. It also lends an air of fantasy, contributing to the *mythos* created.

Common science fiction themes also link *Cosmos* to fiction. Interstellar travel and time travel, two science fiction tropes identified by Alasdair Wilkins,[32] are synonymous with the content of *Cosmos* and occur throughout the series. In the first episode, Sagan travels to the edge of the universe with his audience in the Spaceship of the Imagination. This ship reappears throughout the series, as does the theme of interstellar travel. Sometimes this travel is not just through space, but through time as well. In "Journeys in Space and Time," Sagan travels to Italy to illustrate time dilation at the speed of light. The focus of the illustration are two brothers, Vincenzo and Paolo. One rides a motor scooter for a short time at the speed of light, and the other stays behind. When the traveler returns, he finds his brother an old man while he is the same age as he was when he left. The example is used to explain the special theory of relativity but enters into science fiction when the brothers' fictional story is used to teach it.

The concept of traveling at the "speed of light" is common in science fiction; in this instance, *Cosmos* uses this reference to show what would "really happen" (as Sagan puts it) if people were to travel at that speed. However, the side effects of such travel have not been fully confirmed. In effect he is not reflecting the reality of what "theory" means. Sagan is representing something science *believes* but cannot verify as if it were confirmed truth when he says, "It really happens." Because of his interview in *Rolling Stone*, we can surmise that Sagan saw theories as *both* certain *and* changeable. Since the "myths" of science (theories) may be eventually replaced,[33] just as the myths of ancient Greece were replaced, this distinction becomes important. In *Cosmos,* Sagan uses the authority of Science to present not possibilities but sureties—while personally harboring belief in the fallibility of science and its edicts. The space that our host inhabits, somewhere between absolute truth and a strongly held belief (whatever the source—spiritual, fantastical, or other), is where *mythos* is created.

SCIENCE FICTION ARCHETYPES AND *COSMOS*

Science fiction stories can provide a basis for explicating archetypal structural, visual, or narrative elements. When placed side-by-side, *Cosmos* and *Star Maker*[34] are different artifacts built on similar scaffolding, structured using similar frameworks. The *mythos* created in *Cosmos* because of its similarity to science fiction builds interest within the program, garnering popularity. Comparing a few key components of science fiction with those of *Cosmos*, and then comparing *Cosmos'* overall structure with that of *Star Maker*, can illustrate this point. The following paragraphs first concentrate on the science fiction elements that I've selected for explication. Afterward, subsections describe these elements in *Cosmos*, then *Star Maker*, and finally compare the presentations of science fiction archetypes (or frameworks) between the two.

Genre Archetypes of Science Fiction

Not all science fiction is set in space, but those stories that are will have some key elements in common. Ships in science fiction can be of an infinite variety; however, they all serve similar functions. First, they change the audience's perspective, creating a frame through which they will view the events that follow (*ēthos*). The perspective generated can be either from outside or inside the ship. Either way, ships create "atmosphere" within the program, much like a submarine or battleship may do for a war movie. They also incorporate (and often introduce) technology to inspire the interest and awe of the audience. The ships themselves can elicit wonder since they demonstrate novelty. Of course, they are also a means of survival and critical to the survival of the people inside. If (or when) certain systems break down or get damaged, the beings inside are put at grave risk.

Perspective from outside the ship allows the audience to see differences in size or importance of ships to other elements in the story, for example, the ship to a planet or star, the ship to the universe, or the ship to other ships. The exterior often also provides some clue to the "unique purpose" of the ship: Is it a vessel for exploration? For transport? For defense or aggression? Rarely are black ships considered "good"; this color is often reserved for ships that pose threats—much in the same way white hats were worn by the "good guys" and black hats by the "bad guys" in old movies. Perspective from inside the ship allows the audience to gather the ranking or relative importance of the people inside it based on their accommodations, work stations, uniforms, duties, and the like. It provides a peek at daily life and what comforts may be enjoyed (or discomforts suffered) by the crew. Color and lighting can play an enormous role in establishing this sort of information.

Space is of course closely connected to the ship archetype as one would not exist without the other. As with the ship, space itself is presented as a novelty. Rhetors (writers, directors, producers) will normally attempt to apply some new or unusual angle to space or space travel, however minor. Creating an illusory frame of "zero gravity" in *2001: A Space Odyssey* was achieved by challenging the perspectives of its audience. A spaceflight attendant walks in a circle, from the floor to the wall to the ceiling, and then through a door. A pen floats in the air. Movement is artificially slowed. Walking tracks run in an unbroken circle. All these elements contribute to this fictional frame of space travel. In the science fiction genre, space itself is portrayed as an intriguing, novel place—but it is also portrayed as a place of potential danger and the unknown. The purpose of the focal journey will almost always be related in some way to exploration and discovery (as in the *Star Trek* title monologue). The purpose may also be to solve mysteries. Space can provide a venue for battle. Some inherent element of danger will always exist, and so space itself can be an important element of a program's building of dramatic tension.

Spaceship of the Imagination

Cosmos' means of transport through the universe is depicted as a glowing dandelion seed. Its shape is certainly novel, but it is packed with meaning. It demonstrates that the ship is not one of war, simple transportation, or mere commerce. The ship's shape indicates a purpose connected with life and proliferation, not destruction. This adventure can unlock an understanding of and appreciation for science that Sagan believes can "save" the human race. In this way, even though the ship is not attacked and suffers no equipment malfunctions, it represents the audience's means of survival during the journey. This unusual shape and purpose directs viewers' perspective toward viewing the "travel" as a journey or an adventure, not a mission. Similarly, the interior of the ship provides unspoken information about the "crew," or, in this case, Sagan. He is the only person shown on the ship; a single chair indicates that no other crew members are present nor expected. He works the controls and consults the map, which indicates that he is both our captain and our navigator. He can manipulate the ship to reveal a keyhole "window" in the floor. Like in many science fiction depictions of spacecraft, the environment is clean and white. Also typical of science fiction ships, interior edges are predominantly rounded.

Space is presented as a place of wonder and adventure. In many ways, it is described almost like a Garden of Eden. Sagan frames it as the source for the raw materials make up everything, including people's bodies. The thing that makes humanity different is consciousness—but it is part of space as a whole, essentially the consciousness of space, the way for the universe to "know itself."[35] Physical beings become incorporated into the fabric of the universe and intelligence becomes a functioning part of it. The *mythos* of *Cosmos* illustrates how the human race fits within the larger make up of "all that is or ever was or ever will be."[36]

Star Maker

Olaf Stapledon's 1937 novel *Star Maker* centers on a cosmic journey, undertaken by a lone man who joins a variety of people at different points along the way. Stapledon did not view the book as a work of science fiction; he preferred to view it as "*myth, fantasy, and philosophical fiction.*"[37] Sagan's biographer implies that Sagan may have been a fan of Stapledon's work as a child.[38] If this is the case, it likely influenced his work as an adult. The journeys between *Cosmos* and Stapledon's *Star Maker* certainly mirror each other in several ways, regardless. The "ship"—or manner of interstellar travel—in *Star Maker* is as unique as that in *Cosmos*. The protagonist travels through the universe using his *consciousness alone*, going from one "human" body to another to discover the meaning of life, much like a thought experiment (or Ship of the Imagination). The book describes all intelligent life as "human," no matter what physical form it takes. Some of these species are plant-based; others are sea creatures. No matter what form they take, all intelligent civilizations repeat the same patterns of war and other self-harm. The few that are able to transcend petty differences move forward in their evolution. Those that aren't, are destroyed. Throughout the universe, for all societies, survival is always equated with cooperation. Life itself is of paramount importance. In fact, some civilizations willingly, even joyfully, allow themselves to be destroyed rather than harm another or commit the sin of war.

This mode of travel, although not physically a "ship" made of metal, serves the same archetypal purposes already described. It demonstrates the power of the human consciousness to readers. It allows Stapledon to introduce new perspectives from a variety of species, planets, and times to his readers. The atmosphere creates a novel perspective that combines both being an *observer* with being a *participant*. In some ways, it is almost godlike, as the experiences of the protagonist extend from the beginning to the end of the universe—and even include his observations of earlier universes. The audience learns about the "unknown" and gets the most essential questions of life answered through traveling in this manner.

By traveling through the power of the human mind alone, readers are able to discover and explore from a place of personal safety. The dangers inherent in failing to cooperate are experienced with the emotional pain of witnessing the death and destruction of promising civilizations. Stapledon makes clear the immensity of time and space and how small humans as physical beings are, how short the mortal moment is. Through a space vehicle made up of the human mind, consciousness is elevated as the most powerful—and transportable—trait. The wonder and awe of space common to the science fiction genre is thus formed. By the end of the journey, the protagonist's questions are answered and he is happy to be home again to embrace his "dull" and ordinary human existence. It is through this journey of discovery that the dramatic tension of the story is created.

Comparisons: Star Maker and Cosmos

The similarities between the main thrusts of *Star Maker* and *Cosmos* is striking. *Star Maker* posits that life itself is the most important thing. Even when destroyed, another civilization will rise to take its place, one that is perhaps superior and able to avoid the destruction that claimed the first. Stapledon makes clear in his novel that that science can save and enhance life but it must be balanced and beings must work together in order to really succeed.[39] While the author indicates that religious zealotry is dangerous, he imbues his book with a sort of spiritual quest for the "Star Maker," a godlike creator. When the narrator starts the journey, he cannot see the value or beauty of the "domestic," but he learns the cycles of creation and destruction. In "A Note on Magnitude," which appears after "Epilogue: Back to Earth," Stapledon makes clear the value of life: "A living man is worth more than a lifeless galaxy."[40] In the end, "we" become one with the universe, ourselves the Star Maker, similar to the way Sagan presents humanity as a "way for the *Cosmos* to know itself."[41] In both works, humanity and the universe are interconnected; consciousness and life are the most important things.

Sagan's journey through *Cosmos* shares the same type of meandering structure as *Star Maker*. It does not concentrate on a chronological sequence or on tying together a featured scientist with ones previously discussed in the program. Sagan emphasizes that human life is a valuable thing and, once destroyed, it can never be recreated. He promotes science as having "communal" value for everyone on the planet. If all nations worked together, he argues, science could advance civilization in positive ways that working independently (developing weapons technology, for example) will never be able to achieve. Sagan criticizes the "reptilian brain" as well as the human vulnerability to "pseudoscience" and unprovable claims. Yet, *Cosmos* also incorporates elements of spirituality, claiming that human life (specifically, the audience) is a way for the universe to be conscious or self-aware. The unity of all of creation is presented in *Cosmos* as "all that is or ever was or ever will be"[42]—eternal, like a god.

CREATION OF *MYTHOS*

A science fiction framework can help popular science television to become "something more" than an educational program. Achieving resonance with an audience is the first step in this process. Following "formulaic lines" to form the program's "basic pattern and structure," for both the dramatic arch and for character development (which can also incorporate "stereotypes and shared fantasies"), assist in this pursuit.[43] A combination of "reason" and "imagination" can, according to Francis Bacon, create a "fable"[44] or a myth. This process of *mythos*-building finds its source through either the story's narrative or the visual effects. Of course, a combination of both of these will increase the opportunity for creating fantasy themes and forming *mythos*.

Narrative and *Mythos*

Cultural significance is key to creating *mythos*.[45] A crucial first step to becoming "culturally significant" is addressing the contemporary rhetorical, or *kairotic*, moment. Doing so demonstrates cultural fluency on the part of the rhetor. Understanding what is important to the intended audience, and then addressing those concerns in the artifact created, will increase the chances of resonance and thus *mythos*-building: "Mythic characters, images, figures, and narratives (specific stories and plots) are established by social consensus and usage over time, and they *move, engage us* somehow, from within."[46] *Mythos* in some ways is like a social contract of understanding. In return for due reverence to these important aspects of story-telling, a pathway can be created for the artifact to not only *address* its time, but to *become a symbol* of its time—the essence of *mythos*.

Stapledon creates *mythos* in *Star Maker* through creating "propaganda" pieces that made people think or contemplate and somehow change how they viewed life.[47] Published during the Great Depression and between the world wars, the book is a strong argument for peace and cooperation. The "advanced" civilizations in the book make plain Stapledon's assertion that accepting one's own destruction and death (and even accepting it joyfully) is far better than causing destruction and death.[48] Civilizations that do not learn this lesson are doomed to die, no matter what. Only the societies that value cooperation enough to choose their own death over killing others have a chance for very long-term survival. War met with peace—not aggression—results in understanding,[49] but war met with violence destroys civilizations and squanders wealth.[50] Working toward a collective galactic consciousness, peacefully coexisting, is the only worthwhile goal. It is also the only one that can attain any modicum of success, even though at every stage of development this principle seems to be challenged. *Star Maker* was released after World War I but during Hitler's concerning rise to power, so the issues of passivity reflect a desire to actively avoid another war, whatever the cost—including self-destruction.

Cosmos was created during a time of runaway inflation, economic and environmental concerns, and threat of nuclear war. As with *Star Maker*, the underlying message also concerns peace (this time, a response to the Cold War). The underlying goal is again one of understanding and cooperation, this time focused through the embrace of science. For Sagan, these ideals play a major part in evolving humanity past its "reptilian brain" and the base tendencies that stem therefrom. However, he presents scientific investigation and acceptance of scientific ideals *as the only path to* understanding and cooperation.

In addition, just as *Star Maker* builds a "galactic consciousness," Sagan pronounces humanity as the consciousness of the universe. The audience *is* the universe because all of humanity, the audience's physical beings, are made of "star stuff."[51] In addressing the economy and inflation, he notes how beneficial exploration is to the development of societies[52] and how everyone in the society benefits when science and exploration are embraced. He emphasizes concern about the environment by addressing how much destruction would undoubtedly occur through nuclear war. Our Earth, already so abused, would be destroyed. These issues resonated with contemporary audiences and contributed to the program's creation of *mythos*.

These narratives, one based in science fiction and the other in popular science, both address some of the major issues or concerns of their time. While *Star Maker* did not become as intertwined with its contemporary culture as *Cosmos* did, the archetypal narratives struck a resonant chord with some by impacting contemporary concerns. *Cosmos* did the same. That said, one advantage that *Cosmos* had over *Star Maker* was the medium of delivery. Television allowed Sagan to present visual images that might also resonate with audiences along another sensory input stream. The combined resonance of story with visual stimulation increased the program's chances for creating *mythos* but also for becoming culturally significant as an artifact of its own time.

Music and *Mythos*

Sagan narrated *Cosmos*, but stretches of music, with no voiceover, also provided guidance to viewers. While this work will not delve deeply into the rhetorical nature of music or *Cosmos*' musical score, it is worth mentioning that the music-filled gaps between the narration combine with concurrent visual elements to stimulate these senses as the central locus of attention. Although music in the background ensures that the pictures are not shown in silence, the instrumentals also serve to increase or direct emotional response to the information being presented visually. It guides the audience to a peak of tension; the music can illustrate foreboding or represent joy. This auditory accompaniment directs the audience to experience the visuals on the screen in a *particular* way.

The audience may then come away with their own thoughts or contemplations about what they see on the screen, but these conclusions have been guided, to some degree, by the emotional cues provided by the music. These moments of reflection are not as frequent or extended as those in *2001: A Space Odyssey*, but they can have a similar impact. With a musical accompaniment, a program can leave a picture "on a screen long enough for [the audience] to contemplate it, to inhabit it in [their] imaginations . . . [to be inspired with] awe."[53] This "awe" is imprinted on audience imaginations, a lingering souvenir of the rhetorical construction the program.

Visual Effects and *Mythos*: Transcending Truth and Fiction

Just as lengthy, rich textual descriptions of civilizations and other planets enhance a science fiction novel like *Star Maker,* so visual depictions are necessary for a television audience to connect with a program that explores complex scientific information. Some of these explanatory depictions in *Cosmos* are pictures, but most are computer renderings. Because they are *representations* rather than *reflections* of places or things, an element of fictionalization is interwoven with the science. Roland Barthes argues that, "the 'ethic' of the drawing is not the same as that of the photograph."[54] Because the visuals aren't necessarily a *reflection* of an event or a place but instead an "imagining" of what these things *might* look like, the truth of the depiction can be called into question. To place this assertion in context: a drawing of the Loch Ness monster would not be considered convincing "evidence" regarding its existence; a verifiable photograph might be.

A high-quality, real-to-life computer animation could also be convincing—even though it is simply another form of "drawing"—if it were presented by a trusted host who explains how scientists *think* it lives (without clarifying that they can't prove it even exists). This type of presentation might be mistaken for *evidence* of Nessy's existence rather than as an *illustration* regarding the *belief* or *theory* that it exists and how it lives (if it does). This example may seem a bit extreme, but a phenomenon similar to this one occurred with the Animal Planet program *Mermaids: The Body Found* and its sequel, *Mermaids: The New Evidence,* which resulted in some audiences believing the presentation was popular science. Purposely presented in "documentary form,"[55] some audience members came away believing that mermaids were real, in part, perhaps, because actors were identified as scientists. By presenting actors in this way, the program took the privilege of the cumulative scientific *ethos.* Because of that and the apparently "realistic" computer animations, some audiences were persuaded of the *real existence* of mermaids.

Sometimes editorial manipulation can be obvious (in overly dramatic storylines, for example), but other times it is less obvious. The clothing that the host wears, the tone of voice, the word choice—they communicate an *ethos* that encourages audiences to open up and trust. The quality and placement of animations and graphics communicate that the Big Bang may have looked one particular way, when in reality, the picture is a result of suppositions and imagination. Realism in visuals makes audiences assume that what they are seeing is real, even when it's not.[56] The audience knows intellectually that what they are seeing is not an actual picture—a true representation—but that awareness is relegated to subsidiary awareness (as described in chapter 4). Their focus is on the Big Bang—and then the "knowledge" that they have seen how it would have looked, based on the visual depiction.

Some may argue that visual depictions in popular science are accurate because they reflect scientific data. In truth, these visuals are not *reflections* of data—they are *translations* of it. As Burke points out, translations may come close in meaning to the original, but they are not necessarily fully reflective of it. He calls these differences "displacements" which change the essential meaning from the original term. He uses the example of changing "'mother and father' to parents."[57] These translations create displacements within the text. To say "parents" instead of "mother and father" can reflect a meaning that differs from the original. "Parents" denotes the people who raise a child, which may be mother and father (sometimes including stepparents), two mothers or two fathers. On the other hand, "mother and father" usually denotes the biological or legal relationship of two adults to a child. While this translation does not change the intended meaning in very big ways, it *does* change the meaning.

That said, manipulation of a source does not invalidate it as an authentic interpretation—even if the essence of the artifact translated barbarizes the original. Umberto Eco describes how Stravinsky's *The Rite of Spring* is manipulated by Disney's *Fantasia* to tell the story of dinosaurs and to give an account of Earth's history.[58] While this presentation of the text (music) is unlikely to reflect the intended meaning of the original, the music does fit along with this alternate interpretation,[59] likely because of the careful work of the production staff. Disney's presentation, although manipulative in meaning, at the same time gets the audience to focus on particular elements within the music and "fixes the listeners' attention on effective (for Disney's purposes) musical values.[60]

Just because audiences may not realize that the music is being interpreted differently than intended, that doesn't mean the usage is "unethical." This principle works as well for visual effects. The representation may simply be focusing attention on particular elements prioritized by the rhetor—in the same way Disney focused attention to "the timbre, the rhythms"[61] and other elements in the music.

The same principle applies when translating novels to film. As Eco acknowledges, a story may not indicate the type of clothing that a character wears or even if the character is dressed at all. However, in translating a story to film, characters will normally be dressed. If the novel did not indicate a specific style or fashion, the translation from novel to film is not made less valid through the addition of these elements—even though additional meaning may be added that expands on the intent of the original.[62] Eco considers this difference to be manipulation, but he does not vilify it. It doesn't matter if the implicit is made explicit and later the reverse occurs—text is manipulated in the translation,[63] an unavoidable side effect of the process.

Accommodation of scientific data and theories is likely going to experience the same side effects. Popularizations, in particular, are going to go through a process of translation. One example is when colors are used in visual depictions of space. In these instances, the colors are merely "representative colors [which] translate information in invisible light into a picture we can understand, much as text can be translated from one language to another, or symbols might be used in a cryptogram."[64] These colors determine the area of focus for viewers and they also frame the event or object as one of "beauty." If the light—represented via color—is actually part of an *invisible* spectrum, then can these phenomena be "truly" beautiful to the human eye? The addition of an element that heightens the sensory appeal also adds *value* to the element depicted, altering its true nature. The audience will then believe that they know what something looks like because they will have perceived the visuals as "fact" or "reflection" rather than "translation" or "representation."

At a March 2015 meet-and-greet that followed a public presentation, I informally asked Neil deGrasse Tyson, astrophysicist and host of the 2014 reboot of *Cosmos,* about this issue of "false" color. He responded that false color was a bit of a misnomer. He said that the colors do exist, it is just that humans would have to "have really big eyes" to see them. Although he reported that the color is "genuine," I argue that human eyes *as they are* couldn't really see them. Changing color changes the modality or certainty of the image.[65] When pictures have their color increased, they create high modality or high "believability." In effect, this action increases the "realness" of the picture. If color is lowered or removed from a photo, the modality is decreased. The colors represented in the pictures of nebulae often represent the wavelengths of light. When a nebula's modality is shifted through coloration, then scientists (or graphic artists) are increasing their reality for the audience. Because science needs to be accurate only in terms of measurements and not in terms of what can be seen, then the picture is reflecting "what can be known by the means of science."[66] But it is not reflecting what the human eye can see.

While scientists know and understand the presence of light and color is not exclusively determined by human capabilities, lay persons may not understand the finer distinctions in this. That means altered pictures are effectively translated to accommodate human limitations and aesthetic sensibilities. Therefore, an element of myth-making or fantasy has been added. These photos do not show what humans would *really* see from their spaceship windows (or projection screens). However, most people who have seen these pictures likely believe they reflect what they'd *really* see if they were traveling by in their spaceship. In truth, they *are* representations of what we'd see—but *only* if we had "really big eyes."

Other popular science programs in general (not just *Cosmos*) face the same challenges when translating scientific information for lay audiences. A program about dinosaurs may follow a dramatic pattern in order to make them more interesting or relatable, for example. The program will almost certainly contain computer animation and graphics that show the dinosaurs moving and interacting with their environment. The catch with these types of animations is that audiences may not be fully cognizant that the program has taken some artistic license. Writes J. Anthony Blair, visual depictions like this create a situation in which "we don't just imagine the narrative, we 'see' it unfolding before our eyes. Seeing is believing, even if what we are watching is invented, exaggerated, half-truths or lies."[67] Audiences may realize that storylines about baby dinosaurs in danger are totally fictional, but they may inadvertently believe they *know* how the animals walked, interacted, reacted. They are unlikely to fully realize that they've only seen how scientists *postulate* that these animals behaved. The audience may actually begin associating human characteristics and emotions with the creatures because that is how they are presented.

If the audience *feels* for the dinosaurs (fear, pity) then they're more likely to be interested in them—which means *value has been added* to what they are. This sort of fantasy-building, although innocent in nature, also serves a rhetorical purpose. It persuades audiences to watch and also to *care*. It elicits an emotional connection (*pathos*). Dramatizations usually rely, at least in part, on an appeal to emotion. Creating or changing images to be more attractive to the human eye—for example, adding false color to pictures of space—adds interest or beauty. Adding interest or beauty adds value to the thing depicted; it also creates a layer (however thin) of *pathos*. Adding value *changes the nature* of the thing itself. It also changes the frame through which the audience perceives that thing. Therefore, the graphic image is no longer "reflecting" its subject but is instead only "representing" it.

These visual elements create a frame, a forum from which a rhetor gains persuasive power (*ēthos*). These visuals are necessary in a medium like television in order to attract and retain an audience. Polanyi and Burke both argue that frameworks (or "terministic screens") direct perception in a way that creates a particular "truth." Perception of "truth" can be altered depending on the angle or construction of the frame. In the case of *Cosmos*, the use of visual elements depicts the birth of galaxies, the way in which DNA first replicated itself, and the grandeur and beauty of the library of Alexandria. Just as interweaving photographs from the Viking mission with computer renderings would likely not have been differentiated by the audience (as explored in chapter four), so would other illustrations likely pass the audience unquestioned when presented through a similarly constructed framework. Shifts from photographs to re-enactments or computer animations can be easily missed. Because *Cosmos* shows many artist/computer renderings in a way that passes for photographs (or reflections rather than representations), the audience is likely to accept them as real and not question the verifiability of the image depicted. They will likely not question, but merely accept, what is presented on the screen without pondering the "truth" of the images.

Stating a disclaimer as each rendered image is shown would break into the forum—breaking flow,[68] disrupting *ēthos*—and therefore greatly decrease the program's effectiveness. By breaking in with information from "outside" the carefully constructed forum, Sagan would be fracturing the attention of his audience and incorporating doubt into his own argument—which would lessen his *ethos*. The audience, if given time or desire to examine what they are seeing, would likely realize on its own that the visuals are *constructed* and not *captured*. However, when their attention is directed, uninterrupted, through the story narrative and sensory presentation, they are more likely to get caught up in the program and are less likely to critically examine the particulars. The audience is willingly and readily immersing themselves in the "fantasy" that Sagan is presenting—and at the same time taking his argument from a perspective (forum) he has created. However, when their attention is directed, uninterrupted, through the story narrative and sensory presentation, they are more likely to get caught up in the program and are less likely to critically examine the particulars. The audience is willingly and readily immersing themselves in the "fantasy" that Sagan is presenting—and at the same time taking his argument from a perspective (forum) he has created.

Even though *Cosmos* teaches science, the presence of *mythos* makes the science more interesting for (or palatable to) a lay audience. *Mythos* provides an element of "magic" in the science to capture the audience's imagination but it also provides a familiar structure through which the audience learns about the unfamiliar. *Cosmos* was able to create *mythos* by triggering pleasant associations with contemporary popular culture (*kairos*); by presenting a narrator who respected the audience, trusted them, who did not declare the audience's "ignorance" or talk down to them (*ethos*); by presenting a special effects-laden frame or forum through which the audience could view the journey and hear Sagan's arguments (*ēthos*); and by demonstrating genre similarity to science fiction. He created *mythos* by connecting imagination and fantasy with science through incorporating science fiction-like elements. Without the power of *mythos*, the "lessons learned" in *Cosmos* would have eventually faded or been incorporated into common knowledge, their source of popularization forgotten. Now, the program is continually eulogized anew in modern media like YouTube. Its mythopoetic presentation creates added significance for the program, resulting in long-lasting popularity.

NOTES

1. Spyros D. Orfanos, "Mythos and Logos," *Psychoanalytic Dialogues* 16, no. 4 (2006): 483.
2. Michel Leiris, "The Sacred in Everyday Life," in *The College of Sociology, 1937-39*, ed. Denis Hollier (Minneapolis: University of Minnesota Press, 1988), 24.
3. Orfanos, "Mythos and Logos," 484.
4. Jonathan Cott, "The Cosmos: An Interview with Carl Sagan," *Rolling Stone*, 1980, 45.
5. Paolo Vivante, *The Homeric Imagination: A Study of Homer's Poetic Perception of Reality* (Bloomington, IN: Indiana University Press, 1970), 6.
6. James S. Baumlin and Tita French Baumlin, "On the Psychology of the Pisteis:Mapping the Terrains of Mind and Rhetoric," in *Ethos: New Essays in Rhetorical and Critical Theory*, ed. James S Baumlin and Tita French Baumlin (Dallas: Southern Methodist University Press, 1994), 106.
7. Philip Rose, "Philosophy, Myth, and the 'Significance' of Speculative Thought," *Metaphilosophy* 38, no. 5 (2007): 639.
8. Cott, "The Cosmos: An Interview with Carl Sagan," 45.
9. Ibid., 54.
10. Ibid.
11. Ibid.
12. "Science on the Air," *Science News* 118, no. 8 (1980): 115.
13. Ibid.
14. Ibid.
15. Adrian Malone, "Journeys in Space and Time," *Cosmos: A Personal Voyage* (Studio City, CA: Cosmos Studios, 1980).
16. Adrian Malone, "The Backbone of Night," *Cosmos: A Personal Voyage* (Studio City, CA: Cosmos Studios, 1980).
17. Ibid.
18. Adrian Malone, "Blues for a Red Planet," *Cosmos: A Personal Voyage* (Studio City, CA: Cosmos Studios, 1980).
19. Adrian Malone, "Who Speaks for the Earth," *Cosmos: A Personal Voyage* (Studio City, CA: Cosmos Studios, 1980).
20. Adrian Malone, "The Persistence of Memory," *Cosmos: A Personal Voyage* (Studio City, CA: Cosmos Studios, 1980).
21. Malone, "The Backbone of Night"; Malone, "Journeys in Space and Time."
22. Adrian Malone, "Traveller's Tales," *Cosmos: A Personal Voyage* (Studio City, CA: Cosmos Studios, 1980).
23. Malone, "Journeys in Space and Time."
24. Adrian Malone, "One Voice in the Cosmic Fugue," *Cosmos: A Personal Voyage* (Studio City, CA: Cosmos Studios, 1980); Malone, "Journeys in Space and Time"; Malone, "Who Speaks for the Earth."
25. Malone, "The Persistence of Memory."
26. Adrian Malone, "The Edge of Forever," *Cosmos: A Personal Voyage* (Studio City, CA: Cosmos Studios, 1980).

27. Damien Broderick, *Readng by Starlight* (New York: Routledge, 1995), 8.
28. Keay Davidson, *Carl Sagan: A Life* (New York: John Wiley, 1999), 116–17.
29. Ibid., 112.
30. Martin Green, "Writing About Science," *College English* 43, no. 6 (1981): 570, http://www.jstor.org/stable/376994.
31. Thomas M. Lessl, "Science and the Sacred Cosmos: The Ideological Rhetoric of Carl Sagan," *Quarterly Journal of Speech* 71, no. 2 (1985): 175.
32. Alasdair Wilkins, "Ten Tropes You'll Find in Science Fiction Over and Over Again," *io9*, September 2010.
33. Cott, "The Cosmos: An Interview with Carl Sagan," 54.
34. Olaf Stapledon, *Star Maker* (Mineola, NY: Dover, 2008).
35. Adrian Malone, "Shores of the Cosmic Ocean," *Cosmos: A Personal Voyage* (Studio City, CA: Cosmos Studios, 1980).
36. Ibid.
37. Robert M. Philmus, *Visions and Re-Visions: (Re)Constructing Science Fiction* (Liverpool: Liverpool University Press, 2005), 115. Emphasis in the original.
38. Davidson, *Carl Sagan: A Life*, 115.
39. Stapledon, *Star Maker*, 120–21.
40. Ibid., 255.
41. Malone, "Shores of the Cosmic Ocean."
42. Ibid.
43. Jeffrey W. Kirsch, "On a Strategy for Using the Electronic Media to Improve the Public Understanding of Science and Technology," *Science, Technology, and Human Values* 4, no. 27 (1979): 53.
44. Karl R Wallace, *Francis Bacon and the Nature of Man* (Urbana: University of Illinois Press, 1967), 82.
45. William G Doty, "Exploring the Manifold Spheres of Mythos," *Mythosphere* 1, no. 2 (1999): 118.
46. Ibid. Emphasis in the original.
47. Philmus, *Visions and Re-Visions: (Re)Constructing Science Fiction*, 115.
48. Stapledon, *Star Maker*, 117.
49. Ibid., 148.
50. Ibid., 149.
51. Malone, "Shores of the Cosmic Ocean."
52. Malone, "Traveller's Tales."
53. Roger Ebert, "2001: A Space Odyssey," *RogerEbert.com*, 1998.
54. Roland Barthes, *Image, Music, Text*, ed. Stephen Heath (New York: Hill and Wang, 1977), 43.
55. "'Mermaids: The New Evidence': Never-Before-Seen Footage of Mermaids Allegedly (VIDEO)," *The Huffington Post*, May 2013, http://www.huffingtonpost.com/2013/05/27/mermaids-the-new-evidence_n_3343773.html.
56. David A. Kirby, "Science Consultants, Fictional Films, and Scientific Practice," *Social Studies of Science* 33, no. 2 (2003): 239.
57. Kenneth Burke, *Language as Symbolic Action* (Berkeley, CA: University of California Press, 1966), 66.

58. Umberto Eco, *Experiences in Translation*, ed. Alastair McEwen (Toronto: University of Toronto Press, 2001), 120.

59. Ibid.

60. Ibid., 121.

61. Ibid.

62. Ibid., 123.

63. Ibid.

64. Grace Wolf-Chase, "Coloring the Cosmos," *Phi Kappa Phi Forum*, 2011, 5.

65. Gunther Kress and Theo van Leeuwen, *Reading Images: The Grammar of Visual Design* (New York: Routledge, 1996), 164.

66. Ibid., 169.

67. J. Anthony Blair, "The Rhetoric of Visual Arguments," in *Defining Visual Rhetorics*, ed. Charles A Hill and Marguerite Helmers (Mahwah, NJ: Lawrence Erlbaum, 2004), 56.

68. The concept of flow is explored in "The Forum of Television."

Chapter 6

The Rhetorical Art of *Cosmos*

Cosmos tells the story of humanity; its dramatic structure builds to something Baumlin and Baumlin might call the program's "divine revelation."[1] This revelation is meant to resonate with audiences beyond their immediate interaction with the story and impact their attitudes and even actions beyond the program. By selecting universal questions, asked since the dawn of humanity,[2] Sagan directly ties the *Cosmos* journey with ancient yearnings of the human soul. He helps the audience to wonder about, put visuals to, and contextualize concepts that are normally associated with the spiritual. By connecting with these ancient archetypes, Sagan creates fantasy themes that build an effective and enticing rhetorical vision. The symbolic reality he creates is one in which humanity has the power to control its evolutionary destiny.

The poetic language and stunning visual images are intended to move the audience toward a "soaring spiritual high,"[3] but vignettes about missed opportunities or setbacks in scientific advancement elicit frustration and disappointment. Viewers are meant to regret the budget cuts, which removed the "wolf trap"[4] from the Viking mission to Mars.[5] Sagan's stories about Ionian stagnation,[6] suppression of the dodecahedron,[7] and destruction of the Alexandrian library[8] create wistful glances at opportunity nearly seized.

In *Cosmos'* symbolic reality, the audience becomes one with God and the universe. Inside them is a "little universe,"[9] even as they live within a larger cosmos. Not only are they part of the fabric of creation, at the end of the final episode, Sagan declares that they embody the "eyes and ears and thoughts and feelings of the cosmos."[10] Because of them, the cosmos is not just sentient; it is also self-aware: "star stuff contemplating the stars."[11] The audience is "more than" equal to the stars: they are consubstantial with them. They are unified with their divine guide, Sagan, and everything the universe contains. They are sanctified for their discipleship.

The rhetorical construction of *Cosmos: A Personal Voyage* is so effective that the program did not need time to "build" an audience. It enjoyed immediate popularity. The first two episodes attracted "perhaps as many as 10 million [viewers] each" on their initial run.[12] PBS knew right away that it had a hit on its hands. Given the numerous re-runs of the program, the availability of program DVDs, and the 2014 *Cosmos* reboot that was made nearly thirty-four years after the original, it is clear that Sagan was able to capture the imagination of his audience. In much the same way that the *Star Wars* franchise has been able to maintain a rabid fan base and shape the entertainment that followed, so has *Cosmos* been able to retain its stature as an apex moment in popular science television.

Tapping into the *kairotic* code of its time allowed *Cosmos* to anticipate what viewers would find appealing. It demonstrated an understanding of the "language" that resonated with audiences of the time by referencing *Star Wars* and other popular entertainment. The special effects connected to the cultural milieu, the touchstones of the age. It spoke to concerns about nuclear war, the economy, and the environment. The messages and the way they were presented resonated with the audience. References to fantasy and offering an affordable means of escape in a difficult economic time created an opportunity for audience building. Through it, *Cosmos* gained the audience's trust and esteem.

While addressing its rhetorical situation was key in attracting and retaining its audience at the time it first aired, programmatic elements that were created as a result of *Cosmos*' *kairotic* moment can also be credited with its lasting popularity. This book separately describes connecting to that *kairotic* moment, forming the popular science *ethos*, the creation of *ēthos*, and the resulting *mythos*. The conscious production decisions made in response to the public opinions about science and scientists, the environment and preservation of the planet, and the prevailing forms of popular entertainment in the late 1970s shaped the program's popularity and impact. Its persuasive power is embedded within its rhetorical construction; that rhetorical construction is shaped by a deft ability to craft *Cosmos*' presentation with the interests and concerns of its contemporary audience at the forefront.

Sagan's unimposing persona is also key to the success of the program. His host and guide effectively establishes *ethos* from the first episode. He is at once able to communicate unflagging expertise and knowledge while portraying an unintimidating "everyman," something that helped to bring the public of 1980 into the complex and sometimes intimidating world of science. By introducing spiritual and artistic elements along with the science, Sagan creates a definitive persona that helps his subject matter reach a broader audience who is receptive to his message. Sagan was savvy enough to create a likeable host who does not intimidate the audience with his pedigree as a scientist. Instead, he illustrates great knowledge through his poetic words and references.

The host's persona communicates interest only in *sharing* this vast storehouse of knowledge and the awesome journey he is about to take. Through his movement, Sagan invites the audience to join him on a walk, which eventually becomes this journey through the *Cosmos*. His ship is powerful, but it contains no sign of weaponry, indicating that even if the audience gives him power over themselves, he will pose no threat. In fact, he gives the audience power by showing his own vulnerability: by standing on the edge of the cliff and by showing himself in an imperfect manner with the wind whipping his hair, shadows on his face, eyes squinting. He reveals the importance of the audience as the "consciousness" of the *Cosmos* and does not openly deny any spirituality or religious beliefs—beliefs that might put up barriers to viewers' acceptance of his messages. Instead, he expands the notion of spirituality to include the vast universe at large, with humanity woven into its fabric as its soul. He asks the audience to "skip a step" in their ideas about creation and apply their notions about god to the universe.[13] This persona is able to at once be a guide and a friend who is powerful yet unthreatening, smart, and trustworthy.

As a member of the scientific orthodoxy, Sagan suffered the consequences of his "unorthodox" manner of communicating science, yet his ability to popularize science for millions of people stretches across decades of time. The audience's trust of him and their interest in the subject matter depends on his representation of persona. This persona allowed Sagan to gain *ethos* with a broad 1980 audience who was wary of science and associated it with "pork," weapons of war, and elitism. Unlike orthodox science, which can itself benefit from the formation of persona in its communications, popular science *must* present a host who can break through a vast array of potential barriers, some of which are not connected to science. Sagan works around barriers of belief, varying education levels, geographic and cultural identities, and distrust of his field. Without his host's persona, *Cosmos* would have seemed to be a lecture and not an adventure that people eagerly took: an adventure that people are still taking nearly forty years later.

In *Cosmos*, Sagan displays exceptional ability at building effective frames through which the audience receives his argument. The audience has to trust the speaker in order to enter into these frames because when they do, they are conferring onto the speaker great power. These frames open audiences mentally and emotionally for persuasion by framing the unfamiliar within familiar frameworks, potentially creating a sense of security mixed with adventure. Without the interruptions of commercial breaks (that break flow), Sagan was able to create a forum that allowed the audience to immerse themselves in the voyage.

The visual effects provide interest but also create a connection to science fiction films. These illusions are highly believable and easily absorbed when audiences are experiencing a state of deep flow. These visual effects foster interest but also create for the audience "proof" of assertions and claims, even when the things presented are really imaginings or representations. The Spaceship of the Imagination allows the audience a sense of contribution, the impression of power with responsibility. Emotional manipulation and the mirror effect can further create belief in the experience and program content—even that which is purely imaginary. All these things can be created through *ēthos*.

The combination of these different rhetorical elements in popular science television can lead to the creation of fantasy themes and a science fiction-like *mythos*. By creating *mythos*, by connecting to science fiction as a genre, the cosmos is formed in audience minds in part through education and in part through fantasy. *Cosmos* provides entertainment and has made a lasting impact on popular culture—while teaching science. It gives audiences a way to learn while still dwelling within their imagination. Although some illustrations and re-enactments blur the line between reflection and representation, the program is not fictionalized, but transcends black and white lines. Visual elements create a science fiction-like atmosphere for the program that moves it beyond "popular science" into a hybrid with science fiction, creating a popularizing mythology out of science education.

FANTASY THEMES AND RHETORICAL VISION: SOME NOTES ON *COSMOS*

The rhetorical vision of *Cosmos* centers on the idea of humanity being in control of its own evolution. Beginning from the first episode, Sagan proclaims that "we are made of star-stuff," a statement he repeats several times throughout the series. This seemingly simple phrase connects the audience with the immensity of the universe. It makes them part of its fabric. They are both *part* of the cosmos and the *result of* it. Individual humans have become sentient beings, born out of inanimate matter produced by the stars. Their evolution as a species has taken them from the status of "animals" to "star explorers." The universe, all it contains, all life and matter, are consubstantial. Humans and the stars are one.

Sagan declares that stars "sparked the origin of life." Elements are pushed out into the universe and they eventually form other things, sometimes living things. Some elements eventually became people. To emphasize human importance within the universe, Sagan proclaims in the first episode, "Shores of the Cosmic Ocean," "We are a way for the universe to know itself." This statement is repeated several times throughout the program. Sagan distinguishes the audience as the consciousness of the *Cosmos* and therefore an integral part of all that is.

This sentiment is also present in spirit in episode seven, "The Backbone of Night," when he declares that exploration of the universe is a mission of self-discovery. To emphasize the interconnection of all things, he tells a school child, "*You* are part of the Milky Way."[14] Humanity's "oneness" with the universe is tied to its evolution from "star-stuff" to homo sapiens to the explorers humans will someday become.[15] Humans and the human race begin as elements, born out of the death of a star; evolve as a potential-filled, ever-improving form of life; become a consciousness which learns about the universe and its origins; and eventually returns to the *Cosmos*. Evolution is the natural and appropriate progression. Without it, humans would remain as dust.

Sagan's predominate use of the first-person plural ("we") denotes a shared experience.[16] It communicates the shared nature of the journey, even though the program was filmed months before and individual audience members may be sitting alone in their living rooms. This language indicates that viewers are not alone in their separate homes, but they are traveling together through the universe. Much like the shared nostalgic experiences of attending school together or serving together in the military, the audience has together discovered the universe, the cosmos.

Earth is the sacred home of life. It is the reason to avoid war: to preserve life. It is the starting place for the next step in evolution. However, Sagan warns, humans have the power to "force the hand of nature" and to make Earth "uninhabitable."[17] Our greenhouse gases can turn the relative Eden of Earth into a hellishly hot Venus.[18] Humans could very well destroy not just human life, but all life on Earth. Humanity will make the choice of life or death for the entire planet.[19] He repeats the phrase, "If we do not destroy ourselves"[20] to point out that our still-evolving weakness could result in catastrophic destruction if we do not take control of it. Life must be preserved in order to evolve. Sagan says our extinction would be an act of "unforgivable neglect," but it is not inevitable.[21]

Sagan is the divine leader. He stands as a savior guide who can help the audience make the right decisions to avoid destruction. Free will allows people to have control but Sagan serves as an oracle to show them "the way." According to Sagan's vision, the human species possesses nearly limitless "untapped potential."[22] In "Heaven and Hell," Sagan walks among a model of the planets. His size in comparison to the models is enormous. His "travel" between the planets is easy; his size demonstrates his power. His physicality is presented as human, so audience members can compare their own size among the planets with Sagan's. When he says, "We are star stuff which has taken destiny into its own hands," he shares his power and enormity with the audience. Divinity and power, all the knowledge of the universe and control of time and space, belong to the audience when they follow Sagan.

Evolutionary Destiny

Sagan asserts that humans must take control of their own evolution. Because of the power it gives us, Sagan presents evolution as positive, even "uplifting."[23] He describes it as an opportunity to choose our actions as a species, as inhabitants of this planet and as part of the *Cosmos* that surrounds us. Taking charge of personal evolution is demonstrated through the story of Milton Humason. He was a mule team driver who took charge of his own destiny to become a great astronomer. As a mule team driver, he must have been physically strong; becoming an astronomer indicates he was also intellectually powerful.[24] The framing of Humason seems to indicate that Earth is simply the beginning for the human race—that is, if the audience chooses to set aside its reptilian predilection for war. The spaceship setting symbolizes peace and peaceful exploration. Space is vast and offers technological advances and riches, just like the seafaring voyages of centuries ago did. Humanity must travel to the stars in order to continue evolving.

In "One Voice in the Cosmic Fugue," Sagan notes that selection (natural or artificial) can be imposed from outside. Radiation from the stars impacts evolution-driving mutations.[25] Creation is an "expression" from the "mind of God."[26] However, change is inevitable. Whether it is chosen or inflicted, change will happen.[27] Even though selection can come from outside, it can also be controlled through intelligence. Since the universe and God become the same in *Cosmos*—evoking the same questions about the origin of life and creation[28]—the universe itself is responsible for creation. If humans are consubstantial with the universe, then they have the power to control or influence creation. What they choose to do will determine their own survival or destruction.

In the same episode, Sagan illustrates how humans can guide evolution when he tells the story of Japanese crabs on whose shells appear markings that look like a human face. Around these crabs has grown a myth that the faces are those of Heike warriors, so fisherman will throw the crabs back to the sea when they find them in their nets. Sagan points out that humans directed this evolution—they chose to preserve the lives of the crabs that looked like warriors and to take the lives of others. In the same way, the narrative implies, humans can choose its own destiny and direct its own evolution, just as it did with the crabs. Evolution is a choice. It can be controlled and is already being controlled by human hands.

Cosmos implicitly connects the European enlightenment and exploration. Holland's "flying boats" explored Earth's oceans in the seventeenth and eighteenth centuries and brought back knowledge and social advancements. Their investigations of the planet led to a prosperous time: "Never before or since has Holland boasted such a galaxy of scientists, mathematicians, philosophers, and artists."[29] In "Traveller's Tales," Sagan declares, "Voyageur spacecraft are their descendants." These boats are the ancestors of ships that explore the stars. They have evolved so humanity must follow suit.

Sagan's anecdote about his childhood neighborhood illustrates how limiting staying on Earth would be. Even though neighborhoods beyond his own were unknown to him as a young boy, in order to grow into adulthood, he moved beyond the boundaries of Bensonhurst. It is a natural progression for children to explore beyond their homes. Not to do so would limit their potential. Sagan uses this story as a way of illustrating how space exploration will help "deprovincialize" our planet.[30] Staying confined to the boundaries of Earth would be like children staying in their home neighborhood for their entire lives, never venturing out to discover what else life may have to offer.[31] Their vision would be narrow and their potential for learning, discovery, and wonder stunted.

Sagan warns that humans can't physically adapt fast enough to keep pace with a shifting environment.[32] Human intellect will help the species to adapt and to survive. This control over the effects of environment on the body hints toward human power to change inwardly to evolve outwardly. Journeys to the stars will only be possible if the "reptilian brain"[33] is conquered, if humans renounce aggressive tendencies and their baser instincts or emotions. Embracing intellect will allow humanity to embrace and shape their next step in evolution.

Evolution is necessary and it is already happening. Humans can choose the next stage of evolution through ignorance—through war and destruction, for example. Or, evolution can present a teleological ladder the species can climb to achieve its full potential. When Sagan shows the line drawings illustrating human evolution, beginning with single-celled organisms,[34] he stops with the present human form. However, the implication throughout the series is that humans are, as yet, not fully evolved as a species. His reference to the "reptilian brain" makes that clear. Sagan shows that even the constellations change and evolve,[35] so humans would be no different. We are made of star-stuff. Our origin story, according to Sagan and *Cosmos*, starts and ends with the stars.

Realization and Vision

Preserving life by avoiding war and decrying its weapons will allow the exploration of space to take place. This exploration is necessary to continue on the path toward evolution. Humans can intelligently direct this evolution through supporting science. Continuing to be motivated by "superstition or greed or stupidity" could lead to the destruction of humanity. It is human destiny to travel to the stars: "If we do not destroy ourselves, we will, one day, venture to the stars"[36] The human race must let go of war and selfishness to shape its own fate and evolve to the next stage of development. Without wise choices, evolution will be delayed. Or, potentially, the species may be extinguished.

The path to preservation and advancement will stem from "our compassion and our intelligence." It is the way of superior understanding and indicates an advanced level of intellectual and emotional evolution. Those who follow this path will become the saviors who, through their actions, will make possible "a meaningful life for every inhabitant of this planet." It is also the way "to the stars," and will allow those who work toward this goal to partake in the adventure of space exploration.

To Sagan, space provides the means of human survival and of evolution. These voyages must be driven by discovery and curiosity. Sagan's Spaceship of the Imagination is a vehicle of peace. It represents the commitment humanity will have to make toward peace and cooperation. It has no weapons and no creature comforts. Space travel will not be easy; evolution is a process. The power of salvation has been bestowed on the audience who can now decide the fate and "destiny of our descendants."

In *Cosmos'* rhetorical vision, humans are already evolving past the animal state. They possess planetary consciousness, a collective memory, built through technology. This technology and consciousness can be harnessed to advance the species through exploration. While some scientific advances have been exploited for war, moving beyond these primitive concepts can help humans embrace their "mammalian brain." In order to do that, Sagan proclaims that the first step to evolution is "a fundamental restructuring of economic, political, social, and religious institutions."[37]

Cosmos tells its audience that the human race has an "obligation to survive and to flourish."[38] It isn't enough just to live; humanity is meant to prosper, to evolve. As the consciousness of the universe, the act of exploration is both self-interested and selfless. Humanity will be discovering and learning for the sake of the cosmos itself. According to Sagan and *Cosmos*, evolution is the only way that the human race will be able to sufficiently pay homage to the universe, of which it is a part. His commandment is that humanity values the Earth, supports science, strives toward the next stage in evolution—and ventures to the stars.

FINAL THOUGHTS

The voyage on which Sagan invited his audience on September 28, 1980, has remained (for the most part) unchanged across decades of time. Looking at it through the perspective of the cosmic calendar, thirty-seven years is an eye blink between installments. Looking at it through the perspective of the scientific, entertainment, and technological advances of the last thirty-seven years, this feat is remarkable, at the least. One thing is certainly clear: *Cosmos: A Personal Voyage* is a carefully crafted work of rhetorical art.

NOTES

1. James S. Baumlin and Tita French Baumlin, "On the Psychology of the Pisteis:Mapping the Terrains of Mind and Rhetoric," in *Ethos: New Essays in Rhetorical and Critical Theory*, ed. James S. Baumlin and Tita French Baumlin (Dallas: Southern Methodist University Press, 1994), 106.
2. Adrian Malone, "The Edge of Forever," *Cosmos: A Personal Voyage* (Studio City, CA: Cosmos Studios, 1980).
3. "Shores of the Cosmic Ocean." Ann Druyan introduction of year 2000 re-issue.
4. So the opportunity to find evidence of microbial life on the planet was lost.
5. Adrian Malone, "Blues for a Red Planet," *Cosmos: A Personal Voyage* (Studio City, CA: Cosmos Studios, 1980).
6. "Journeys in Space and Time"
7. "The Backbone of Night"
8. "Shores of the Cosmic Ocean"
9. "One Voice in the Cosmic Fugue"
10. "Who speaks for the Earth?"
11. "Who speaks for the Earth?"
12. Frederick Golden, "The Cosmic Explorer," *Time*, October 1980, 63.
13. Malone, "The Edge of Forever."
14. Adrian Malone, "The Backbone of Night," *Cosmos: A Personal Voyage* (Studio City, CA: Cosmos Studios, 1980).
15. If humans do not first "destroy themselves" before they have the opportunity.
16. Ernest G. Bormann, "Fantasy and Rhetorical Vision: The Rhetorical Criticism of Social Reality," *Quarterly Journal of Speech* 58, no. 4 (1972): 402, doi:10.1080/00335637209383138.
17. Adrian Malone, "Heaven and Hell," *Cosmos: A Personal Voyage* (Studio City, CA: Cosmos Studios, 1980).
18. Ibid.
19. Adrian Malone, "Who Speaks for the Earth," *Cosmos: A Personal Voyage* (Studio City, CA: Cosmos Studios, 1980).
20. Adrian Malone, "Journeys in Space and Time," *Cosmos: A Personal Voyage* (Studio City, CA: Cosmos Studios, 1980); Malone, "The Backbone of Night."
21. Adrian Malone, "Encyclopedia Galactica," *Cosmos: A Personal Voyage* (Studio City, CA: Cosmos Studios, 1980).
22. Adrian Malone, "One Voice in the Cosmic Fugue," *Cosmos: A Personal Voyage* (Studio City, CA: Cosmos Studios, 1980).
23. Malone, "Blues for a Red Planet."

24. Malone, "The Edge of Forever."

25. Adrian Malone, "The Lives of the Stars," *Cosmos: A Personal Voyage* (Studio City, CA: Cosmos Studios, 1980).

26. Adrian Malone, "The Harmony of the Worlds," *Cosmos: A Personal Voyage* (Studio City, CA: Cosmos Studios, 1980).

27. Malone, "Heaven and Hell."

28. Malone, "The Edge of Forever."

29. Adrian Malone, "Traveller's Tales," *Cosmos: A Personal Voyage* (Studio City, CA: Cosmos Studios, 1980).

30. Malone, "The Backbone of Night."

31. Ibid.

32. Adrian Malone, "The Persistence of Memory," *Cosmos: A Personal Voyage* (Studio City, CA: Cosmos Studios, 1980).

33. Malone, "The Backbone of Night"; Malone, "The Persistence of Memory."

34. Malone, "One Voice in the Cosmic Fugue"; Malone, "Journeys in Space and Time"; Malone, "Who Speaks for the Earth."

35. Malone, "The Edge of Forever."

36. Malone, "The Backbone of Night"; Malone, "Journeys in Space and Time."

37. Malone, "Who Speaks for the Earth." Sagan is specifically addressing the potential for nuclear war in this scene. However, these words apply to human evolution in general as well.

38. Ibid.

Chapter 7

Cosmos for Another Generation

Cosmos: A Spacetime Odyssey (*Cosmos* 2014) aired from March to June 2014, introducing the brand to a new generation of viewers. Sponsored by Samsung Galaxy and Chrysler,[1] the program received a large-scale launch, with episode one enjoying a "multinetwork simulcast" on ten separate FOX networks.[2] Although Ann Druyan, an original cowriter, had first started thinking about "reviving" the brand "in earnest" in 2007 (when, she says, "public hostility towards science was really naked"), it wasn't until she found a network willing to give her creative control that the *Cosmos* revival materialized.[3]

Druyan wanted to retain control in order to preserve the *Cosmos* legacy, believing (understandably) that she understood the program better than anyone else could.[4] Seth MacFarlane (of *Family Guy* fame) as producer, Neil deGrasse Tyson as host, and Ann Druyan with Steven Soter as writers may have seemed like an unlikely team at first, but this group provided the program with some distinct advantages. MacFarlane's "street cred" with boundary-pushing entertainment on both television and in movies and Tyson's reputation as a scientist and popularizer spoke of a modern "edutainment" experience. With Sagan gone, the Druyan/Soter team was as close to reuniting the original scribes as would have been possible.

Sagan's original *Cosmos* had the benefit of airing on PBS, a channel with a reputation for both education and for quality entertainment, so audiences in 1980 were likely confident that the program would be interesting and entertaining. *Cosmos* 2014 didn't necessarily have that advantage. FOX's reputation as a network, particularly as a network that produces educational programs, is tenuous at best. The FOX News branch of the network is largely criticized for having a highly conservative bent; the FOX entertainment network has been known for bawdy humor (*Married with Children*, *The Simpsons*) since its launch. Despite whatever challenges, however, the team put together a quality program: the original series won three Emmy awards (two for creative technical crafts and one for program planning), but *Cosmos* 2014 won four (two for music, one for sound, and one for writing).

How did the reboot follow in the tradition established by the original? In what ways did it break new ground? This chapter will attempt to answer these questions by examining *Cosmos: A Spacetime Odyssey* using the same rhetorical lenses used in earlier chapters of this book. Although it is still too early to make definitive conclusions in regard to things like program longevity, it is worth considering the ways the program both paid homage to Sagan and the original program while establishing its own rhetorical voice. While this chapter will not attempt to go into the same level of depth as earlier chapters, it will provide a general overview of *Cosmos* 2014's fantasy-infused *mythos* using a structure that closely mirrors the rest of the book. The chapter organization will follow the book's structure as a whole.

VIEWER RECEPTION AND *KAIROS*

According to Tyson, *Cosmos: A Spacetime Odyssey* was shown in 181 countries and 47 languages. This wide distribution is likely boosted by its prestigious pedigree. Journalist Hank Campbell notes that the program was actually the "4th of 4" in ratings when it aired but it "tied in the key 18–49 demographic rating and share." Still its audience of "3.5 million"[5] was impressive for a popular science program. Campbell's choice to refer to viewers as "devout believers" indicates he does not believe they were seeking entertainment alone. He likely expected viewers to have already been fans of Sagan, *Cosmos* and/or Tyson.

Another journalist summarized *Cosmos* 2014's reception this way: "Despite ecstatic reviews, 'Cosmos' (sic) has generated a relatively modest audience[6] of about 4 million viewers in a tough Sunday time slot."[7] However, the program has "performed well with the notoriously difficult-to-reach young male cohort."[8] Re-airings, DVD/Blu-ray sales, streaming numbers, and production of additional installments will determine to what extent the public embraces this program.

Regardless of motivations or intentions going into the program, some of the information it presents is certainly controversial to some. Creationists took exception to the presentation of evolution; they even tried demanding "equal time" on primetime to explore their theories.[9] According to Michael Gryboski, *Cosmos* 2014 was controversial from the start, covering "culture war issues like [evolution] and the history of science in western civilization." One Oklahoma affiliate "accidentally" interrupted the first-episode's discussion of evolution by breaking in with a station promo.[10] The program stirred up debate but, according to Hank Campbell, it also provided arguments to refute climate-change denials.

Weapons, War, Science, and Scientists

War is not presented as the major threat of human extinction in this installment of *Cosmos*. In this round, *Cosmos* 2014 instead focuses its gaze on climate change and the repercussions of it. The topic itself seems to be controversial despite the overwhelming support of the idea from the majority of the world's scientists. This angle of controversy is fueled by distrust of the scientists themselves. Contemporary critics of this theory often point to the fact that scientists make their money by doing science and making claims. In essence, some of the arguments claim that scientists are motivated by money, not the good of the public. These claims echo the distrust of the 1970s.

Interestingly, *Cosmos* 2014 addresses this issue of "scientists on the payroll" but frames the scientists denying climate change as the ones being "bought." In episode 7, "The Clean Room,"[11] Tyson tells the story of Clair "Pat" Patterson, the scientist who fought for years to get lead out of gasoline, arguing that it was harmful to humans. The featured scientist on the other side of the argument is Robert Kehoe, who the program portrays as a man hired to "cloak the threat." The episode sets up Patterson as a scientist hero and Kehoe (unfairly or not) as a scientist villain. During one segment, Patterson miraculously arrives "on the fifth day of testimony" to testify at a hearing about lead levels in the environment. This hearing was scheduled to be held when Patterson was supposed to be away in Antarctica. Tyson hints at a tinge of "conspiracy" in that inconvenient timing.

It took twenty years, according to Tyson, but lead was eventually banned. Improvement in human lead levels was nearly immediate and now, he says, the scientific consensus is "unanimous" on the subject. The parallels to climate change, government inaction, and scientists hired by corporations with vested interests in carbon-based fuels seem to be clear. The program implies that once climate change is finally addressed (like it or not by those who are against it), the scientists who are currently denying it, perhaps no longer enticed by the income offered via groups lobbying against its veracity, will again step back in line with the majority consensus.

It's an interesting twist to defend science by indicting scientists. The method for figuring out which scientists are "bad" seems, in this scenario, to rely heavily on figuring out which ones are "outliers." If one scientist goes against the majority, that scientist should be singled out for skeptical assessment and motive analysis. Even though many segments throughout both of the *Cosmos* installments illustrate that scientists going against the grain (like

Giordano Bruno) were correct in their assertions, in this one way, the program seems to say that the outliers are not to be trusted.

The Economy

The costs associated with science and space exploration did not seem to be as much of a concern for the new installment of the series. Perhaps the increased visibility of external trappings of affluence would not have been as out of place in this installment. If the show had been produced during the height of the mortgage crisis, the Wall Street financial bailout, and unemployment woes of the beginning of this century, perhaps concerns similar to those expressed in 1980 would be present. A deeper investigation into the ways in which public tastes for designer labels and expensive automobiles (if present) are expressed in the design of the ship and the clothing of the host may be warranted.

The Environment

Whereas nuclear war is framed as the greatest threat to human life on planet Earth in Sagan's original *Cosmos*, in *Cosmos* 2014, the greatest threat posed is global climate change. Tyson mentions the topic in no fewer than five episodes; in several others, he alludes to it with stories (like about Clair Patterson in episode 7). He first mentions global climate change in episode 2 by equating the current debate with the "uproar" Darwin caused with his ideas.[12] He waits until episode 6 to bring it up again; this time, it is in relation to plants and their vulnerability.[13]

By episode 9,[14] Tyson's climate change warnings are gaining urgency. He brings the program back to the "Halls of Extinction," a conical pyramid he has introduced earlier in the series. Toward the end of the episode, he equates climate change with the asteroid that killed the dinosaurs: "The dinosaurs never saw that asteroid coming . . . . What's our excuse?"[15] Two episodes later, he describes how an abrupt shift in climate wiped out the great Mayan civilization.[16] Finally the argument reaches its peak in episode 12, "The World Set Free," with the entire episode focusing on the effects of greenhouse gasses, the greenhouse effect, and climate change. Tyson also explains the difference between climate and weather. It is the crescendo of a deftly and slowly built knock-out punch.

Entertainment

The special effects are beautifully done. In many cases, the presentations are so artistically rendered that any number of screenshots for even a single episode could be made into posters or art prints. Today's entertainment options include such a vast variety of options—and entertainment screens are so large with such high definition—that creating an experience that incorporates beauty and that triggers the senses seems mandatory. Sagan's *Cosmos* already set the bar for cutting-edge visual effects, so its progeny would certainly have been expected to follow suit.

The shooting locations, for the most part, seem to stay somewhat closer to home than the original. With the capability clearly displayed by the graphics crew, additional set visits likely would have been completely extraneous. Rhetorically, a decision to stay "closer to home" may have been made consciously as a way to underscore the immediacy of the climate change problem—that it isn't happening "elsewhere," but it is happening "here," in the home country/city in which individual viewers are watching the program. This decision also echoes the first episode's initial voyage, which starts at Earth and moves outward.

NEIL DEGRASSE TYSON'S COSMIC *ETHOS*

Neil deGrasse Tyson is a credentialed scientist, just as Sagan was. Also like Sagan, he had already appeared on television, "written bestselling books" and enjoyed "quite a fan club"[17] before *Cosmos* 2014. About a year after the program aired, Tyson received the same award from the National Academy of Sciences that Sagan had won: the Public Welfare Medal.

Even though he was not an "unknown" before the program, Tyson's public profile seems to have grown since *Cosmos* 2014. He has over six million followers on Twitter and is now host of a new talk show.

A good portion of the success of *Cosmos* 2014 rested on the shoulders of the new "face" of the *Cosmos* brand, Neil deGrasse Tyson. Taking over a program from the person who was as intertwined with the program as Sagan had been certainly must have been daunting. Sagan's name has been synonymous with *Cosmos* for over three decades. Despite this challenge, Tyson was able to pick up the "baton" from Sagan and move the program into the new century, in part because of the introduction of the program and the purposeful manner in which homage and respect was paid to its former leader.

Passing the Baton

Episode 1 of *Cosmos* 2014, "Standing Up in the Milky Way," was critical for establishing the host's credibility, just as it was for the original program. Not present on the streaming or DVD versions of the program is Barack Obama's introduction, in which the then-president tips his hat both to the program and to Sagan. Then, the program starts with a black, blank screen and a familiar voice: Sagan saying, "The universe is all that is or ever was or ever will be." The first image shown is an enormous wave crashing in slow motion on the rocks. This image changes to one from the original program: a dandelion seed pinched between Sagan's fingers. The audience cannot see anything but Sagan's hand, but for those who are familiar with the original, it may well be clear *whose fingers* they are. "Come with me," the voiceover commands as the fingers let go and the dandelion flies through the air toward Tyson. Tyson is standing on a familiar cliff—the same one in Monterey, California, on which Sagan stood to invite us on the original cosmic journey.

Soon after this brief opening and Tyson's initial "introduction," he is shown walking along the cliff, from the left to the right—just as Sagan was shown, all those years ago. He is walking and talking, with the camera showing him from a distance and then moving toward him. The immediate hat tip paid to Sagan and the original *Cosmos* is likely intended to display an "unthreatening" attitude to his fans. The display has no feel of "copying" or straight "duplication"—its approach is one of respect and remembrance. Tyson is not there to take Sagan's place; his goal is merely to continue the journey. He is walking the path out of respect, not as a replacement. He also says, "Come with me," asking us to join him on the journey. It is an invitation, not a command.

Although episode 1 does not attempt to replicate the original, it does follow certain patterns (with a few alterations) and references it several times. The images of the ocean return, but they are not so heavily emphasized. The program sets off early on a journey, but whereas Sagan takes the audience to the end of the universe and then brings them closer to Earth, Tyson begins at the Earth and then his journey extends outward.

A space-capable vehicle is still our manner of transport, but with many differences (as explored below). One of the differences is that Sagan's ship is called the Spaceship of the Imagination. For Tyson, it is merely the Ship of the Imagination. Perhaps the vernacular was changed to make the vehicle seem more updated; perhaps the ship was renamed because Tyson's travels in places that aren't necessarily in "space" (one is inside of a bear, for example). Another of the signature effects of the original, the cosmic calendar, is also repeated. Although updated visually to reflect current special effects capabilities, the calendar has remained very similar to the one Sagan presents. As in the original *Cosmos*, this calendar returns through several episodes.

In the original program, Sagan uses first person plural almost exclusively. He does not mention his own name nor does he list his vast and impressive credentials. In *Cosmos* 2014, Tyson does not say his own name, either, with one exception: in episode 1, he reads his first name from an inscription Sagan wrote in a book he gave the young Tyson: "To Neil, a future astronomer."

Tyson remains fairly anonymous outside of this demonstration of his connection to Sagan, but he repeatedly mentions Sagan's name throughout the program. In fact, he refers to Sagan most often as simply, "Carl," denoting familiarity rather than formality. In this episode, Tyson also lists some of Sagan's notable accomplishments. Highlighting the achievements and contributions of his predecessor demonstrates that he acknowledges his own place in whatever hierarchy might have been established in viewers' minds. Sagan was first: he is the sage, the teacher. Tyson is the one who comes after: the disciple, the student-turned-teacher.

Episode 1 ends with Tyson telling a heartfelt story of meeting Sagan "on a snowy day" in December 1975. He then credits Sagan for inspiring him. He reports that his career plan was already set on being a scientist. Sagan inspired our new host by providing an example of "kind of person [he] wanted to become." The generosity and genuine esteem of this declaration cements Tyson's admiration for Sagan and shows him to be not a threat, but rather an ally, someone who was fortunate enough as a young man to meet Sagan. The reverence of the story and the life-shaping impact of this meeting on Tyson indicate that he is also, at heart, a fan. Following Tyson will not be a betrayal to Sagan; instead, it will be continuing in the legacy Sagan left behind.

Setting

Cosmos 2014 was set, once again, in many places throughout space and time. Two of the most noticeable changes in the setting were Tyson's dress and his ship. Whereas Sagan's dress is nice enough (although limited) and clearly professorial (yet casual) throughout the program, Tyson's dress is a bit more varied, including both more clothing and a greater range of clothing. The wardrobe included a blue chambray shirt with white buttons, light blue shirt with tortoise shell buttons, blue striped shirt, black dress shirt (with a placket to hide the buttons), a black (or deep indigo) t-shirt, a blue t-shirt, mid-thigh-length trench coat, charcoal jacket, black blazer, navy blazer, dusty green anorak jacket, black dress shoes, brown (or tan) shoes that were sometimes dusty, black sneakers, dark jeans, black dress pants, and gray pants. Amazingly, this list is not exhaustive, but illustrates the variety. He occasionally wore his shirt untucked; sometimes his sleeves were rolled.

The variety in his dress may be greater than Sagan's but the color palette—although different—is similarly limited. Tyson's clothing features black, grey, and blue hues whereas Sagan wears orange, khaki, and burgundy. Perhaps the "updated" dress could be said to feature more traditionally "masculine" colors. Fashion in the 1970s tended to be more colorful and vibrant. Overall, Tyson's clothing is a little sleeker and more stylish. Perhaps this styling is a way to avoid the "scientist nerd" stereotype from contemporary popular entertainment (like *The Big Bang Theory*). Tyson still wears no lab coat to distinguish himself as a scientist, but he does wear a "clean suit" in episode 7, "The Clean Room." This suit serves a purpose to show Tyson immersed in the science. For the remainder of the program, his dress does not distinguish him as a scientist.

Although Tyson's clothing varies, he only wears one outfit (with one barely perceptible exception) when he is in the Ship of the Imagination. In these scenes, he wears his sleekest, most expensive-looking, and most "dressy" clothes: black dress shirt (with placket), black dress pants, black blazer, and black dress shoes. Only in the very first episode, when the ship is first introduced, does Tyson spend even a few seconds inside the ship wearing "street clothes." He is standing on the cliffs at Monterey, then the camera pulls out and the audience sees he is now standing inside the ship, wearing the same blue denim shirt and charcoal jacket. The ship pulls away and the next time Tyson is shown, he is already in his black outfit. The impression is somewhat *Matrix*-like, as the characters in that film imagined their own clothing for their forays into the matrix. They chose all black, usually leather. While Tyson is not wearing a leather jacket, the impression of sleek style is still present. One interesting fact and potentially unrelated note about this imagery is that the director of photography for *Cosmos* 2014 was Bill Pope, "whose credits include the *Matrix* trilogy."[18]

Tyson's ship is sleeker, polished, and solid black. Sagan's ship glows, perhaps is even iridescent, and features sharp spindles emanating from a center nucleus. Tyson's ship, on the other hand, is smoother, more curved rather than sharp. It also boasts a center nucleus from which extends a flatter, rounded top and bottom. The outside of the ship doesn't emit light; instead it reflects its surroundings. Whereas Sagan's ship is an obvious "translation" of a dandelion seed, no such obvious translation exists for Tyson's ship. In some ways, it is reminiscent of a star with an elliptical orbit around it. During the opening montage, it seems that a connection between the ship and the human eye might be intended. Regardless, "The new ship is sleek, minimalist and positively gleaming."[19] Tyson described it in an interview as "badass."[20] It certainly presents some of the most interesting visuals of the show, since it reflects its surroundings so well that it almost becomes "cloaked," as in science fiction programs and film.

One additional aspect of the ship that is of particular interest: Tyson may command the ship—but in this new installment of the *Cosmos* series, he also shares it. Tyson inhabits the space with a hologram of DNA in episode 2, "Some of the Things that Molecules Do," John Herschel's picture in episode 4, "A Sky Full of Ghosts," and Wolfgang Pauli—who is portrayed by an actor—in episode 6, "Deeper, Deeper, Deeper Still." Instead of simply referring to events and places of interest *outside* the ship, Tyson brings items and people of interest *into* the ship several times. Instead of simply being pointed to as a "sight" to see, the act of bringing the imagery into the ship symbolizes it being brought into our imaginations. The distinction may indicate a sense of interaction, of invitation not just to see and to learn but to contemplate, even deeper, within our individual imaginations.

Movement

Tyson's movement on the Monterey coast is highly reminiscent of Sagan's (although not identical). He does not stand with his back to the camera there, but he does so on several other occasions. In episode 12, "The World Set Free," the camera comes from behind Tyson as he is speaking. His hand movements are not toward us, but toward the window or projection screen—against which he is silhouetted. Then, the camera glides around to show him from the front. This sort of positioning and the camera angle—emphasizing the audience's movement toward him—could make it seem as though they are *outside* the story. Or, it could make it seem that he is together with the audience on the tour, looking out the same window, rather than serving as a "tour guide" who may miss the sights because he is facing backwards.

In addition, Tyson "controls" the ship, but no panel with buttons or other obvious navigational equipment exists. The lack of controls indicates that the ship is steered via imagination alone. It also allows *Cosmos* 2014 to avoid a potential pitfall of "dating" the program. While Sagan's controls are "modern" for 1980 and reminiscent of *Superman: The Movie*, they still date the program. Simply hovering and moving his hand over them to control the ship does not illustrate well the use of imagination itself as the guiding power. Tyson's ship, with its complete absence of controls, more clearly communicates this intention.

Another difference in movement involves the ship's multiple "windows." Sagan's ship has a television-shaped window at the front and a keyhole-shaped window in the floor that can open at a single gesture from Sagan. In Tyson's Ship of the Imagination, two windows (in addition to the front screen) provide us with a way to see the universe. One of the round windows is on the floor, the other is on the ceiling. The former is for looking at the past, according to Tyson. The latter is for looking into the future. These windows provide our host with additional places to occupy within the ship; he is not limited to the front nor to the back. He can stand in the center and gesture upwards or downwards. The variation provides more opportunities for blocking scenes and thus increasing visual interest. The two additional windows also provide potential as useful points of reference for camera movement "into" a segment.

ĒTHOS: BREAKING THE SPELL

Another difference between the two series installments: *Cosmos* 2014 appeared on commercial television and so needed to include about fifteen minutes of commercials per episode. The commercials interrupt the building of storylines and therefore regularly break the "flow" of the forum. Episodes average four commercial breaks each, spaced around ten to twelve minutes apart. At times, Tyson merely says, "Come with me." The program then goes to commercial, so the audience *cannot* follow Tyson until it returns. These teasers interrupt the story *before* the commercial actually occurs.

The program uses other sorts of "cliff hangers" or "teasers" designed to encourage audiences to return after the commercial break. In episode one, "Standing Up in the Milky Way," one teaser informs the audience that Giordiano Bruno suffered "the most vicious form of cruel and unusual punishment,"[21] then cuts to a commercial without specifying what the punishment was or why he was punished. Tyson does not immediately fulfill the "promise" of the tease when the program returns, either. Since intriguing information is sometimes pulled out of the narrative to encourage the audience to return after a break, the line of argumentation is not always presented chronologically. As Nasser Zakariya points out, "Lead-ins to the advertisements . . . tend in the second series to be dramatized as moments of high tension or expectation."[22] Building tension that is then interrupted by commercials—and not resolved until minutes after the program returns—may cause additional difficulties to the audience achieving flow.[23] It can also be jarring for audiences.

Of course, because of these teasers, the opportunity for achieving flow is broken for all presentations of the program, whether they occur on commercial television or not. The teasers *remain* in the DVD/Blu-ray and streaming versions of the program. These teasers create remnant seams, which interrupt the narrative thread of the storylines. They also make obvious that the program was created for television—which will undoubtedly make viewers, even ones who are deeply engaged, aware of under what circumstances they are watching the show. They are likely to be brought out of the narrative with the intentional "break" caused by these teasers or cliffhangers. When the forum of engagement, *ēthos*, is interrupted, audiences are pulled out of the program into their "real" surroundings. This break may mean that the frame through which arguments are presented is not as strong as it could have been since achieving flow would be much more difficult—and would only be possible in ten to twelve (rather than forty to fifty) minute spurts.

MODERN *MYTHOS*

The opening montage for *Cosmos* 2014 is very different from that of the original. Lasting about a minute, the camera travels first through a valley with barren mountains rising around it. This trip is deeply reminiscent of Sagan's trip through the Mariner Valley in his episode 5, "Blues for a Red Planet." The camera then goes up and peers down at a ragged crater. Inside the crater, galaxies are reflected. In a moment of special effects magic, the crater almost seamlessly becomes a human eye. The ship is reflected in it, then the ship appears over a moon (or planet) with the sun rising. The ship's capability to reflect its surroundings is revealed, then the Milky Way. Eventually this scene fades into a large seashell on a beach, which is picked up by a child. The shell, held up to the sun, fades into a spinning galaxy. A dandelion's seeds are blown into the wind—and one seed transforms into a Voyageur probe. The remaining images include a DNA strand, simple form of life, a blue nebula, an explosion (perhaps a supernova), and then a highly stylized version of the cat's eye nebula. The ship flies into the pupil and the title replaces it: "*Cosmos*," then "*A Spacetime Odyssey*."

Through the "human" elements and references, this montage clearly communicates a connection between people and the universe. Just as Sagan pronounced that "we are made of star stuff,"[24] so does this montage indicate that we are in the universe and the universe is in us. The investigation that is implied seems to focus on closer examination of space, planets, and ourselves, rather than the emphasis on the far reaches of space—as was present in the original's montage.

Re-enactments occur throughout *Cosmos: A Spacetime Odyssey*, as they did in *Cosmos: A Personal Voyage*. However, this time they are usually animated rather than "costume dramas,"[25] creating what one critic called "'graphic novel' interludes."[26] This translation of historical events into animation was likely meant to appeal especially to teenaged and young adult fans of comic books and graphic novels. This change may also have been part of an attempt to limit some of the inherent kitsch of actor-based re-enactments.

Ultimately, the only way to determine the depth of any *mythos* created by *Cosmos: A Spacetime Odyssey* remains to be seen. The masterful address of *kairos*, *ethos*, and *ēthos* certainly can result in *mythos*. When *mythos* is present, so are they, but there is no "sure-fire" recipe. Popularity (repeats, sequels, sales/rentals, views) and the spawning (or inspiration) of fan homages and fan culture over time will become the ultimate litmus test. Although Tyson's profile seems to have increased since the program, it is nearly impossible to say if there is a lasting mark on the culture because of the show. Catch phrases from the show being repeated in other genres and comedian send-ups of Tyson would be one barometer that indicates a chaining out of a fantasy theme. Perhaps a re-airing of the show, or another installment, may inspire these. Sagan had the benefit of a smaller pool of competition, but he also had the benefit of a signature way of speaking, something Tyson lacks.

The producers and writers had a much bigger challenge for *Cosmos* 2014 than they did in 1980. Then, they were breaking new ground. With the new installment, they were planting in long-cultivated earth. Somehow, they needed to make a program that fulfilled all the promises of the *Cosmos* brand, made an honest stab at creating its own *mythos*, and yet stayed close enough to the original to avoid cries of "heresy" from the public. Audiences now are much more accustomed to visual effects and much wiser to leading arguments, as well.

For now, the public reception seems generally positive and ratings were high enough to prompt discussion/conjecture about another installment in the series (unconfirmed). To Tyson and team's credit, they did not attempt to replace, duplicate, or even necessarily "update" the juggernaut original. In doing so, they would have inadvertently implied that *Cosmos: A Personal Voyage* was no longer relevant. Instead, this installment *builds* on the original, continuing the legacy. It serves as a testimony to Carl Sagan and his exceptional talent as a science popularizer: likely one of Druyan's key goals in seeking a rebirth of the brand. *Cosmos: A Spacetime Odyssey* also has introduced Tyson to very large audience. His brand may one day be as big or even bigger than Sagan's—but only time will tell.

NOTES

1. "Fox Networks Group Announces First-Ever Simultaneous Cross-Network Global Premiere Event for 'Cosmos: A Spacetime Odyssey,'" *PR Newswire*, 2014.
2. Ibid.
3. Meredith Blake, "'Cosmos' Co-Creator Ann Druyan Looks Back in Wonder on Her Odyssey," *LA Times*, June 2014.
4. Dave Itzkoff, "'Family Guy' Guy as Astrobiology Guy," *The New York Times*, February 2014.
5. Hank Campbell, "Cosmos Ratings in the Home Stretch," *Science 2.0*, 2014, http://www.science20.com/science_20/blog/cosmos_ratings_in_the_home_stretch-137670.
6. "Modest" here must be taken in comparison to other primetime shows, not necessarily other popular science programs.
7. Blake, "'Cosmos' Co-Creator Ann Druyan Looks Back in Wonder on Her Odyssey."

8. Ibid.

9. "Creationists Demand Airtime on Neil deGrasse Tyson's 'Cosmos,'" *The Huffington Post*, March 2014.

10. Ibid.

11. Brannon Braga, Ann Druyan, and Bill Pope, "The Clean Room," *Cosmos: A Spacetime Odyssey* (Studio City, CA: Cosmos Studios, 2014).

12. Brannon Braga, Ann Druyan, and Bill Pope, "Some of the Things that Molecules Do," *Cosmos: A Spacetime Odyssey* (Studio City, CA: Cosmos Studios, 2014).

13. Brannon Braga, Ann Druyan, and Bill Pope, "A Sky Fill of Ghosts," *Cosmos: A Spacetime Odyssey* (Studio City, CA: Cosmos Studios, 2014).

14. Brannon Braga, Ann Druyan, and Bill Pope, "The Lost Worlds of Planet Earth," *Cosmos: A Spacetime Odyssey* (Studio City, CA: Cosmos Studios, 2014).

15. Ibid.

16. Brannon Braga, Ann Druyan, and Bill Pope, "The Immortals," *Cosmos: A Spacetime Odyssey* (Studio City, CA: Cosmos Studios, 2014).

17. Nicola Davis, "Cosmos: How the Creator of Family Guy Remade Carl Sagan's Pivotal TV Series," *The Guardian*, 2014.

18. Ibid.

19. Ibid.

20. Ibid.

21. Brannon Braga, Ann Druyan, and Bill Pope, "Standing Up in the Milky Way," *Cosmos: A Spacetime Odyssey* (Studio City, CA: Cosmos Studios, 2014).

22. Nasser Zakariya, "Exhibiting Cosmos," *Technology and Culture* 56, no. 3 (July 2015): 741.

23. The concept of flow is explored in greater depth in chapter 4 of this book.

24. Adrian Malone, "Shores of the Cosmic Ocean," *Cosmos: A Personal Voyage* (Studio City, CA: Cosmos Studios, 1980).

25. Zakariya, "Exhibiting Cosmos," 742.

26. Davis, "Cosmos: How the Creator of Family Guy Remade Carl Sagan's Pivotal TV Series."

Appendix 1

Cosmos: A Personal Voyage Scene Index

Times listed in this index are rounded to the whole minute. Time markers may vary depending on medium (e.g., streaming vs. DVD). As a result, check within two minutes before and two minutes after the minute-marker estimate given. This index is meant to provide a general guideline only.

Table A1.1 Episode 1, "The Shores of the Cosmic Ocean"

Minute	*Notable Content*
6	"We are made of star stuff."
14	We have grown to consciousness.
20	"Are they also a danger to themselves?"
20	Stellar Nursery.
23	Refers to the "family of the sun," personifying the universe. Kuiper belt. Valles Marineris on Mars.
26	Valleys of Mars.
29	Sagan is off the ship, green grass; tree. Shows a montage from different cultures.
31	Circumference of the earth figured out by Eratosthenes. Discusses Library of Alexandria.
40	Says if he could go back in time, he would go back to the Library of Alexandria.
42	Defines the word "cosmos."
43	Ptolemy.
44	Hypatia's martyrdom mentioned but not explained or described.
47	Aristarchus.
50	Johannes Kepler.
52	Cosmic Calendar.

Table A1.2 Episode 2, "One Voice in the Cosmic Fugue"

Minute	*Notable Content*
2	Organic chemistry.
4	War first mentioned—connected to human-directed evolution. Emperor Antoku. Heike Warriors.
10	Artificial selection.
17	Cosmic Calendar. DNA.
20	Garden of Eden.
22	Cambrian Explosion.
25	Line drawings of evolution.
31	40-second review of line drawings.
34	Symbiotic relationship with plants explained.
37	Molecules, proteins, DNA.
38	Blood cell.
47	Cornell has tried to recreate life.
49	Scientist in background.
50	Bishun Khare.
54	EE Salpeter.

Table A1.3 Episode 3, "The Harmony of the Worlds"

Minute	*Notable Content*
	Kepler. Astrology in newspaper columns.
1	Stellar nurseries. Pleiades, Crab Nebula.
4	Watching people from a rooftop.
12	Constellations.
16	Defines the word "planets." Anasazi people.
21	Ptolemy. Kepler. Claudius. Nicolaus Copernicus.
28	Geometry.
38	Tycho Brahe.
53	The Somnium, science fiction novel written by Kepler.

Table A1.4 Episode 4, "Heaven and Hell"

Minute	*Notable Content*
2	Tungus people and the Tunguska Event.
8	Comet Encke.
11	Edmond Halley and Halley's Comet.
18	Model of solar system.
22	Moon impact; Gervase of Canterbury.
30	Velikovsky.
36	Galileo and Venus.
37	Prism colors; types of light.
41	Venus. Soviet Vanera.
44	Greenhouse effect on Venus.
49	Montage of natural disasters.
55	Global climate. "If we destroy the earth…"

Table A1.5 Episode 5, "Blues for a Red Planet"

Minute	*Notable Content*
	Martians.
2	*The War of the Worlds.*
7	Percival Lowell. Giovanni Schiaparelli.
14	John Carter novels. Alfred Russell Wallace.
17	Robert Goddard.
22	Seeing Earth from the outside, like an alien, looking for life.
26	Mars.
28	Pyramids of Elysium. Olympus Mons. Mariner Valley.
30	Viking probes.
41	Sagan in a lab.
42	Evolution.
50	Wolf Vishniac and Antarctica.

Table A1.6 Episode 6, "Traveller's Tales"

Minute	*Notable Content*
5	Voyager 1 and 2. Io, Europa, Ganymede, Calisto, Jupiter.
8	Golden record.
9	Compares costs and times of space exploration to ocean exploration.
18	John Locke.
24	Antony van Leeuwenhoek.
26	Christiaan Huygens.
39	Europa. Voyager.
45	Io volcano.
51	Saturn.
55	Titan.

Table A1.7 Episode 7, "The Backbone of Night"

Minute	*Notable Content*
1	Sagan's childhood neighborhood in Bensonhurst.
6	Sagan in his sixth grade classroom answering students' questions.
15	Explains how the Milky Way got its name.
16	Rules of nature.
21	Thales.
23	Anaximander.
24	Polycrates. Theodorus.
28	Empedocles. Democritus.
35	Pythagoras.
39	Plato.
45	Copernicus and Aristarchus.
49	In classroom with kids.

Table A1.8 Episode 8, "Journeys in Space and Time"

Minute	*Notable Content*
3	Constellations and how they change over time.
8	Space and time are intertwined.
10	Speed of light.
11	Albert Einstein.
15	Special theory of relativity.
25	Time dilation.
26	Leonardo da Vinci.
28	British Interplanetary Society 1939.
29	Project Orion.
31	Interstellar ramjet.
33	Speed of light.
36	Space travel and time travel are connected. Time paradox. Alternate timelines and dimensions.
40	H.G. Wells' *Time Machine.*
41	Ionians.
49	Line drawings of evolution.
52	Dinosaurs and extinction.

Table A1.9 Episode 9, "The Lives of the Stars"

Minute	*Notable Content*
3	Apple pie and the creation of the universe.
7	Infinity.
8	Googol and Googolplex.
10	Atoms.
20	Nebula.
32	Supernova.
39	Crab Supernova.
41	Neutron star.
44	Gravity.
45	Black holes.
47	Gravity.
53	Stellar nurseries.

Table A1.10 Episode 10, "The Edge of Forever"

Minute	*Notable Content*
4	Bard's galaxies.
6	Laws of physics universal. Galaxy shapes, collisions.
7	Ring galaxy.
8	Quasars.
10	Our sun's orbit around the Milky Way.
11	Doppler Effect/Doppler Shift.
13	Milton Humason. Edwin Hubble was his mentor.
20	Distance of galaxies measured by color shifts.
23	Space is curved. Edwin Abbott—Flatland.
30	Tesseract.
31	Expanding, curved universe.
34	Wormhole.
37	Eastern views of religion and nature.
41	Universe deaths and rebirths.
50	Radio telescopes in New Mexico.

Table A1.11 Episode 11, "The Persistence of Memory"

Minute	*Notable Content*
2	20 Questions and "bits" of information.
9	Regina Maris.
10	Sagan sings whale songs.
10	DNA illustration.
20	Gene library.
25	Reptilian brain.
26	Cerebral cortex.
27	Frontal lobes.
31	Sagan's memories.
35	Thought distinguishes our species.
40	Communal memory.
43	Writing is our greatest invention.
47	Collective knowledge.
51	Extraterrestrial intelligence.

Table A1.12 Episode 12, "Encyclopedia Galactica"

Minute	*Notable Content*
2	Betty and Barney Hill.
4	Hypnosis "reveals" alien abduction.
10	Joseph Fourier. Jean Francois Champollion.
12	Egyptian hieroglyphics.
19	Great Temple of Karnak.
21	Rosetta Stone.
29	Science provides a "cosmic Rosetta Stone."
32	Radio astronomy.
36	Frank Drake; Drake Equation.
45	De-provincialization of planet; search for sounds of life elsewhere.
49	Star wars wouldn't happen because one civilization would be much more advanced than another.

Table A1.13 Episode 13, "Who Speaks for the Earth"

Minute	*Notable Content*
6	Conquistadors sought gold, not knowledge. Brought death.
10	Comes upon "intelligent" planet; communication abruptly cuts off; assumed destruction.
12	Returning to earth. Again, the sound disappears.
16	Reptilian voice. Rejected science's methods.
17	Nuclear explosion with mushroom cloud.
18	Nuclear bombs are "genies of death." Levels of armament.
21	Expense of war.
22	Wonders how we'd explain ourselves to aliens.
23	Urges change with rhetorical questions.
25	Eratosthenes. Library of Alexandria. Aristotle. Ptolemies.
30	"The permanence of the stars was questioned. The justice of slavery was not."
31	Average people weren't involved with science and didn't know about it enough to value it.
32	Hypatia
34	Cyril and mob murdered Hypatia.
35	Destruction of Library of Alexandria.
36	Sophocles' plays.
37	Creation of galaxies.
41	DNA. Line drawing of evolution.
48	Science "connects us with our origins."
49	Science is a tool.
54	Dandelion seed. Lets it go. Ship with empty chair.

Appendix 2

Cosmos: A Spacetime Odyssey Scene Index

Times listed in this index are rounded to the whole minute. Time markers may vary depending on medium (e.g., streaming vs. DVD). As a result, check within two minutes before and two minutes after the minute-marker estimate given. This index is meant to provide a general guideline only.

In the tables, the first time is taken from the network broadcast and includes commercials. The second time is from the 2014 Blu-ray set of *Cosmos: A Spacetime Odyssey*. Please note that this edition of the Blu-ray has errors in naming some of the episodes. These errors have been corrected here. They have been cross-checked and verified.

Table A2.1 Episode 1, "Standing up in the Milky Way"

Broadcast Minute	*DVD Minute*	*Notable Content*
7	7	Viking probe. Mars.
8	8	Jupiter.
9	9	Voyager 1 probe and the golden record.
		COMMERCIAL
14	11	Planets.
16	13	Milky Way.
17	14	Cosmic address.
18	15	Multiverse.
		COMMERCIAL
27	20	Bruno.
29	23	Aristotle.
		COMMERCIAL
33	23	Italy.
36	26	Cross imagery.
38	28	Cosmic Calendar (only image shown).
		COMMERCIAL
41	28	Cosmic Calendar explained.
44	30	Stellar nursery.
50	36	"Last second of the Cosmic Calendar."
		COMMERCIAL
52	36	Back in time on "December 31."
56	40	Sagan's voice. Story about meeting Sagan as a student.
59	43	"Come with me—our journey is just beginning."

Table A2.2 Episode 2, "Some of the Things that Molecules Do"

Broadcast Minute	*DVD Minute*	*Notable Content*
4	4	Dogs.
5	5	Ice age. Breeding dogs and evolution. Animation.
7	7	Ship of the Imagination (not Spaceship of the Imagination).
		COMMERCIAL
11	9	Experiencing a break from the ice age.
12	9	Ship through artery into bear's heart, eggs, DNA strand.
17	14	Natural selection.
18	15	Global warming.
18	16	COMMERCIAL
23	16	Tree of life.
29	22	Resistance to evolution.
		COMMERCIAL
32	23	How vision evolved.
37	28	Evolution.
38		COMMERCIAL
42	28	Halls of Extinction.
46	32	Library of Alexandria
49		COMMERCIAL
52	34	Saturn. Titan.
56	40	History of life on Earth.
58	41	Mentions Carl Sagan. Evolution line drawings compressed into 40 seconds.
59	42	Sagan voiceover.

Table A2.3 Episode 3, "When Knowledge Conquered Fear"

Broadcast Minute	*DVD Minute*	*Notable Content*
2	2	Pattern recognition.
4	4	Defines the word "disaster."
9	9	Jan Oort
10	10	Comet approaching sun.
12		COMMERCIAL
14	13	Edmond Halley.
18	15	Hooke's Law. Robert Hooke.
19	16	Democracy.
22		COMMERCIAL
25	19	Isaac Newton.
29	22	Halley meets Newton.
31		COMMERCIAL
34	25	Halley funds Newton.
40	30	Saturn.
41		COMMERCIAL
44	31	Halley's accomplishments/legacy.
51	37	Halley's prediction.
52		COMMERCIAL
54	38	Halley's comet.

Table A2.4 Episode 4, "A Sky Full of Ghosts"

Broadcast Minute	*DVD Minute*	*Notable Content*
2	1	William Herschel.
5	5	Light bends.
6	6	Light year definition.
7	7	Crab Nebula.
12	12	Beginning of time.
		COMMERCIAL
16	13	Solar system.
17	14	Cosmic Calendar.
22	19	Albert Einstein's insight was "earth shaking."
		COMMERCIAL
26	19	Einstein's theory of relativity.
32	24	Speed of light.
		COMMERCIAL
36	25.30	William Herschel. John Michell.
39	28	New York City.
40	29	Gravity.
41	30	Black hole.
		COMMERCIAL
45	30	Black hole.
50	34	Ship goes into black hole.
		COMMERCIAL
53	35	Black hole.
58	40	Tyson meets Sagan (story).

Table A2.5 Episode 5, "Hiding in the Light"

Broadcast Minute	*DVD Minute*	*Notable Content*
2	2	Space. Nebula.
3	3	Cave people.
4	4	Fire. Motze.
8		COMMERCIAL
11	8	Ibn al-Haytham (Alhazen).
13	10	Concept of zero.
18	15	Space baby.
19		COMMERCIAL
23	16	Walking in a forest.
25	18	Isaac Newton. Light spectrum.
26	19	William Herschel. Infrared light.
29		COMMERCIAL
32	22	Joseph Fraunhofer. Maximilian I.
40	30	Birth of astrophysics.
41		COMMERCIAL
44	31	Field of flowers.
50	37	Atomic signature.
51		COMMERCIAL
53	37	Joseph Fraunhofer.
57	41	Spectral lines.

Table A2.6 Episode 6, "Deeper, Deeper, Deeper Still"

Broadcast Minute	*DVD Minute*	*Notable Content*
2	2	Atoms.
6	6	Plant cells. Stomata.
7		COMMERCIAL
10	8	Plants.
13	10	Climate change.
15	12	Madagascar's rain forests.
17	14	Smells trigger memories.
18		COMMERCIAL
22	16	Inhaling.
25	18	Democracy. Greece. Thales.
26	20	Democritus of Abdura.
30		COMMERCIAL
33	23	Material things don't touch.
37	28	Sun.
40		COMMERCIAL
44	30	Neutrinos.
47	34	Wolfgang Pauli. Red canon ball.
50	37	Wall of Forever.
		COMMERCIAL
53	37	Defines "Wall of Forever."
55	39	Cosmic Calendar.
59	43	Thales and Democritus.

Table A2.7 Episode 7, "The Clean Room"

Broadcast Minute	*DVD Minute*	*Notable Content*
3	3	Creation of the solar system.
9		COMMERCIAL
12	9	Collision course asteroid.
15	12	Radioactive decay. Uranium meteorites.
18		COMMERCIAL
22	15	Zircon crystals help to find age of Earth.
26	18	Sterilized lab.
30	22	Age of Earth.
31		COMMERCIAL
34	23	Saturn.
37	26	Lead poisoning in Rome.
39	27	Lead paint.
40	29	Ethyl Co.
41	30	Science used to hide an environmental threat. Climate change. Robert Kehoe. Clair Patterson.
42		COMMERCIAL
45	31	Clair Patterson.
52		COMMERCIAL
55	37	Clair Patterson. Edmund Muskie. Hearings.

Table A2.8 Episode 8, "Sisters of the Sun"

Broadcast Minute	*DVD Minute*	*Notable Content*
2	2	Devil's Tower. Pattern recognition.
4	4	Pleiades.
8		COMMERCIAL
11	8	Edward Charles Pickering. Annie Jump Cannon. Henrietta Swan Leavitt.
14	10	Spectrum lines.
15	12	Cecelia Payne. Modern astrophysics.
16		COMMERCIAL
20	13	Theoretical and atomic physics.
23	16	Classifying stars.
25		COMMERCIAL
30	18	Episode returns.
34	22	Sun.
36		COMMERCIAL
40	25	Nebula.
42	27	Supernova.
44	28	Orion's belt. Alnilam. Hypernova.
45	29	Backbone of night. Milky Way.
46	31	Aborigines. Supernovas.
47		COMMERCIAL
51	32	Carina Nebula.
52	35	Eta Carinae.
56	38	Light is clean energy.
57	39	"A still more glorious dawn awaits. Not a sunrise, but a galaxy rise."

Table A2.9 Episode 9, "The Lost Worlds of Planet Earth"

Broadcast Minute	DVD Minute	Notable Content
4	4	Lignin.
6	6	Nova Scotia.
8	8	Permian extinction. Hall of Extinction.
10	10	Global warming.
11	11	Methane. Ozone layer.
13		COMMERCIAL
16	13	Guadalupe Mountain Chain.
17	14	Cosmic Calendar.
19	16	Alfred Wegener. Pangea.
22	18	Marie Thorp.
23	20	Autobiography of Earth.
23		COMMERCIAL
27	20	Mariana Trench.
30	22	Bioluminescence.
32	24	Core of Earth.
35		COMMERCIAL
38	28	Planet of the Mammals.
41	30	Sea currents changed and altered global climate.
43	32	Jupiter.
		COMMERCIAL
47	31	Inside Ship of the Imagination.
50	35	Carbon dioxide into environment.
51	37	Halls of Extinction. Unmarked corridor.
52		COMMERCIAL
54	37	Drawings of evolution.
58	41	Pyramid/Halls of Extinction.

Table A2.10 Episode 10, "The Electric Boy"

Broadcast Minute	DVD Minute	Notable Content
3	3	Michael Faraday. Isaac Newton. Albert Einstein.
10		COMMERCIAL
14	11	Humphrey Davy.
21		COMMERCIAL
25	17	Christmas lectures.
30		COMMERCIAL
34	22	Electricity. Magnetism. Light.
40		COMMERCIAL
44	28	Magnetism.
49	33	Aurora Borealis.
50		COMMERCIAL
53	34	James Clerk Maxwell.
56	37	Red canon ball. Pendulum.

Table A2.11 Episode 11, "The Immortals"

Broadcast Minute	*DVD Minute*	*Notable Content*
0	0	Voice over: "Must we die?"
3	3	Enheduanna, Akkadian princess.
5	5	Odysseus. Gilgamesh.
9	9	Sagan voiceover. Mars. 1976 Voyager probes.
10	10	Nakhla meteorite (Martian).
		COMMERCIAL
13	10	Mars.
18	15	Interstellar odyssey.
		COMMERCIAL
22	15	Dandelion.
39	21	Project Diana.
33	26	Life expectancy of civilizations.
		COMMERCIAL
37	26	Enheduanna.
39	27	Abrupt climate change. Supernova.
43		COMMERCIAL
47	32	European invaders of Americas. Bacteria. Microbes.
48	33	Civilizations self-destruct.
50	35	Images of Hitler, Mussolini.
52	37	Cosmic Calendar.
		COMMERCIAL
55	38	Death of our sun.

Table A2.12 Episode 12, "The World Set Free"

Broadcast Minute	*DVD Minute*	*Notable Content*
1	1	Venus.
4	4	Greenhouse effect.
5	5	Soviet Venera.
6	6	White cliffs of Dover.
10	10	Charles David Keeling. Carbon dioxide rates.
		COMMERCIAL
14	10	Ice cores from Greenland and Antarctica.
15	11	Infrared of Earth.
17	13	Volcanoes.
19	14	Purple carbon dioxide.
20		COMMERCIAL
24	16	Earth. Svante Arrhenius. EO Hulburt. GS Callendar. Frank Baxter.
26	18	Carl Sagan's research into greenhouse effect on Venus.
27	19	Difference between weather and climate.
30		COMMERCIAL
34	23	Permafrost is thawing.
38	26	Augustin Mouchot.
40		COMMERCIAL
44	28	Egypt 1913. Frank Shuman.
49		COMMERCIAL
52	33	Nuclear arms race/Cold War. Kennedy speech.
55	35	Atmosphere is a thin layer.
57	37	Venus.
59	39	Kennedy speech.

Table A2.13 Episode 13, "Unafraid of the Dark"

Broadcast Minute	*DVD Minute*	*Notable Content*
		Library of Alexandria. Hypatia. Pythagoras. Euclid.
3	3	Internet is our own "electronic Library of Alexandria."
5	5	Nebula.
8	8	Victor Hess.
10	10	Cosmic Rays.
11		COMMERCIAL
13	11	Fritz Zwicky. Neutron stars. Pulsars.
15	13	Gravity warps the space around it. Light.
17	14	Vera Rubin.
18	15	Fritz Zwicky. Dark Matter. "What we know of the universe is but foam on the waves."
19		COMMERCIAL
23	17	Dark Energy.
26	29	Standard candles. Edward Hubble. Galaxies expanding.
28		COMMERCIAL
31		Program returns from commercial.
32	22	Ocean of space. Voyager 1 and 2.
35	24	Heliosphere.
37	27	Longest phase of Voyager mission.
		COMMERCIAL
40	27	Carl Sagan.
41	28	Golden record. Math and physics as the universal language.
44	31	Cosmic Calendar.
		COMMERCIAL
47	31	One month equals one billion years.
49	33	Sagan member of Voyager imaging team. Directed picture of Earth from Neptune.
54	38	Five simple rules.
58	42	Sacred searching.

Appendix 3

Terms and Definitions

Below you will find definitions for key terms used throughout this book. The definitions are not meant to provide an exhaustive account of all available definitions. Instead, these definitions are meant to provide an explanation for how they will be used in the specific context of this book. Other researchers in other contexts may emphasize different aspects or even redefine certain elements of these terms.

Accommodation or Popularization: often used as interchangeable terms. Both accommodations and popularizations are concerned with making specialized and often highly complex information more understandable for lay (nonexpert) audiences. However, popularizations tend to have broader and larger audiences. They will often cross boundaries with entertainment genres as part of its persuasive appeal. Accommodations stay somewhat closer to the original scientific text. Genres of accommodation may include grant applications or extension brochures. Popularizations are usually created for mass media (news stories, magazines, television). Both types include scientific ideas, theories, facts, investigations or the like that have gone through a process of "translation" or adaptation to make the information more easily understood by a target audience outside the sphere of professional scientists. However, people at all levels of education and income are exposed to popularized science on a nearly daily basis, whether that science is focused on medicine, exercise, research/experimentation/investigation, astronomic discoveries, weather, or the like.

Archetype: recognizable patterns of action, location, character types, traits, or plot themes. In modern terms, they may be considered "formulas" because they use familiar plots and characters.

Artifact: sometimes called a "case." An artifact is the material of focus for rhetorical analysis. The artifact examined in this book is *Cosmos: A Personal*

Voyage. It will also provide the focus of investigation and argumentation. An artifact can also be called a "text."

Audience: the people who receive and possibly act on (or benefit from) a message. If a program is a product, the audience is its consumer. It is made up of the people the program wants to attract, whether the goal is to achieve particular demographics/ratings or for persuasive intent.

Contemporary Audience: refers to the original, primary audience for an artifact. A contemporary audience can also include the *target audience* (which may or may not have been reached). *Cosmos*'s contemporary audience would have been the people who watched its initial run in the Fall of 1980. However, its *audience* includes anyone who has watched the program, no matter when it was viewed.

Cosmos: abbreviated title of the original television series, hosted by Carl Sagan (*Cosmos: A Personal Voyage*). In the chapter of this book focusing on the 2014 soft reboot hosted by Neil deGrasse Tyson (*Cosmos: A Spacetime Odyssey*), the newer installment of the program will be abbreviated as *Cosmos* 2014.

Ethos: one of the rhetorical appeals, the others being *logos* (logic) and *pathos* (emotion). *Ethos* concerns the credibility or reliability of the speaker. It must be earned by the speaker and given by the audience. How the text is presented as well as the content of the presentation contribute to the speaker's *ethos*. Inaccurate information and sloppy appearance will not gain the audience's confidence and trust in the same way accurate information and professional appearance will. Speakers establish authority when they establish credibility. The focus of this book's exploration of *ethos* in *Cosmos* will be on Carl Sagan as the narrator and guide (host) rather than on the production itself. *Ethos* is explored with more depth in "*Ethos* and the Popular Science Persona."

Ēthos: from the same Greek root as *ethos* (*ήθος*). While it is also a way for the speaker to establish authority, *ēthos* is concerned with the frame or *forum* of presentation. *Ēthos* does not need to refer specifically to a concrete physical location (like a living room); a forum can be created virtually through special effects and imagery. *Ēthos* is explored with more depth in "The Forum of Television."

Fantasy Theme: setting, character, and action fantasy themes work together to establish a rhetorical vision.

Kairos: rhetorical concept about capturing or creating the right moment for a message or artifact. While *chronos* focuses on time, *kairos* refers to opportunity. *Kairos* is explored in more depth in "*Cosmos'* Context in Time."

Lay Audience: nonexperts and nonprofessionals who likely have an interest in the accommodated subject matter presented. A lay audience may not understand the technicalities of the research or the history of the findings when presented in scientific terms, but they have an interest in the subject matter. They normally require some "translation" from professional scientific language and format in order to grasp the information presented. Popular science is usually intended for a lay audience. In popular science, readers need to "feel included in a conversation rather than preached at or talked down to."[1] Sarah Tinker Perrault dubs this attitude "positive politeness."[2] She also divides popular science approaches into three categories: "strategies for putting readers at ease, for inviting the reader into the science, and for creating inclusive 'we' that positions readers and writers together in thinking about science-related issues."[3] While these approaches are about presentation, they are customized for a particular type of audience: a lay (or nonexpert) audience.

Mythos: rhetorical concept that is created when an artifact taps into these universal/ancient themes (or myths) in a way that causes a lasting resonance with the audience. Myths are tales that were used to explain truths about human origins, existence, and frailties. They were not created as intentionally fictional but rather as entertaining explanations to answer big questions or to provide origin stories. The concept of *mythos* is explored in greater depth in chapter 5.

Persona: projection of character or false characterization of personality. Persona is a "front," the intended (and often fictionalized) display of personality. Persona can be displayed through dress, mannerisms, word choice, vocal inflection, emotion, reaction, or any other quality that assists in establishing an identity (personality) of the character for an audience. Persona is explored in chapter 3.

Popularization: see "Accommodation."

Rhetor or Speaker: the person responsible for the construction and method of delivery of an artifact. Rhetors need not consider themselves rhetoricians; speakers need not speak (they may also write, for example). Although *Cosmos: A Personal Journey* boasts three credited writers, Carl Sagan is generally given primary credit for shaping the program. He will be referred to as the speaker or rhetor of *Cosmos*, although I here acknowledge that he did not work alone.

Rhetorical Vision or symbolic reality: the combined impact of fantasy themes. It is the "shared symbolic ground exhibited by a vision's participants."[4] It is an understanding of subject or artifact that is shared and often appreciated in the same way by a particular group of people.

Science: refers to "orthodox" or "normal" science. In other words, it refers to science as a profession or a field of inquiry. When referring to orthodox science as a type of "establishment" or intellectual collective, this word can be written with an uppercase "S."

Speaker: see "Rhetor."

Symbolic Cue: these cues are similar to "inside jokes"[5] in that they are understandable only by people who have shared similar experiences or who share particular information. They indicate "membership" in a rhetorical community. By referencing a single symbolic cue, complex, unstated meaning can be readily understood by that community.

Symbolic Reality: see “Rhetorical Vision.”

Text: general term meant to refer to any exchange, object, person, video, writing (or other communication event) that can be analyzed or examined. Texts are not confined merely to the written word. For example, a movie can be considered a “text.” The text of focus becomes a rhetorical analysis’s *artifact*.

NOTES

1. Sarah Tinker Perrault, *Communicating Popular Science* (New York: Palgrave Macmillan, 2013), 105.
2. Ibid.
3. Ibid., 106.
4. Ernest G. Bormann, John F. Cragan, and Donald C. Shields, “Three Decades of Developing, Grounding, and Using Symbolic Convergence Theory (SCT),” in *Communication Yearbook 25*, ed. William B. Gudykunst (Mahwah, NJ: Lawrence Erlbaum, 2001), 273.
5. Ernest G. Bormann, “Fantasy and Rhetorical Vision: The Rhetorical Criticism of Social Reality,” *Quarterly Journal of Speech* 58, no. 4 (1972): 396–407, doi:10.1080/00335637209383138.

References

Badash, Lawrence. "Nuclear Winter: Scientists in the Political Arena." *Physics in Perspective* 3, no. 1 (2001): 76–105.

Barthes, Roland. *Image, Music, Text.* Edited by Stephen Heath. New York: Hill and Wang, 1977.

Bassett, Samuel E. "Homeric Criticism." *The Sewanee Review* 30, no. 4 (1922): 390–98.

Baumlin, James S. "Introduction." In *Ethos: New Essays in Rhetorical and Critical Theory*, edited by James S Baumlin and Tita French Baumlin, xi–xxxi. Dallas: Southern Methodist University Press, 1994.

Baumlin, James S., and Tita French Baumlin. "On the Psychology of the Pisteis:Mapping the Terrains of Mind and Rhetoric." In *Ethos: New Essays in Rhetorical and Critical Theory*, edited by James S Baumlin and Tita French Baumlin, 91–112. Dallas: Southern Methodist University Press, 1994.

Bitzer, Lloyd F. "The Rhetorical Situation." *Philosophy and Rhetoric* 1, no. 1 (1968): 1–14.

Blair, J. Anthony. "The Rhetoric of Visual Arguments." In *Defining Visual Rhetorics*, edited by Charles A Hill and Marguerite Helmers, 41–61. Mahwah, NJ: Lawrence Erlbaum, 2004.

Blake, Meredith. "'Cosmos' Co-Creator Ann Druyan Looks Back in Wonder on Her Odyssey." *LA Times*, June 2014.

Bormann, Ernest G. "Fantasy and Rhetorical Vision: Ten Years Later." *Quarterly Journal of Speech* 68 (1982): 288–305. doi:0.1080/00335638209383614.

———. "Fantasy and Rhetorical Vision: The Rhetorical Criticism of Social Reality." *Quarterly Journal of Speech* 58, no. 4 (1972): 396–407. doi:10.1080/00335637209383138.

———. "Symbolic Convergence Theory: A Communication Formulation." *Journal of Communication* 35, no. 4 (December 1985): 128–38.

Bormann, Ernest G, John F Cragan, and Donald C Shields. "In Defense of Symbolic Convergence Theory: A Look at the Theory and Its Criticisms After Two Decades." *Communication Theory* 4, no. 4 (1994): 259–94.

———. "Three Decades of Developing, Grounding, and Using Symbolic Convergence Theory (SCT)." In *Communication Yearbook 25*, edited by William B Gudykunst, 271–313. Mahwah, NJ: Lawrence Erlbaum, 2001.

Bormann, Ernest G., Becky Swanson Kroll, Kathleen Watters, and Douglas Mcfarland. "Rhetorical Visions of Committed Voters: Fantasy Theme Analysis of a Large Sample Survey." *Critical Studies in Mass Communication* 1 (1984): 287–310. doi:10.1080/15295038409360038.

Braga, Brannon, Ann Druyan, and Bill Pope. "The Clean Room." *Cosmos: A Spacetime Odyssey*. Studio City, CA: Cosmos Studios, 2014.

———. "Deeper, Deeper, Deeper Still." *Cosmos: A Spacetime Odyssey*. Studio City, CA: Cosmos Studios, 2014.

———. "The Electric Boy." *Cosmos: A Spacetime Odyssey*. Studio City, CA: Cosmos Studios, 2014.

———. "Hiding in the Light." *Cosmos: A Spacetime Odyssey*. Studio City, CA: Cosmos Studios, 2014.

———. "The Immortals." *Cosmos: A Spacetime Odyssey*. Studio City, CA: Cosmos Studios, 2014.

———. "The Lost Worlds of Planet Earth." *Cosmos: A Spacetime Odyssey*. Studio City, CA: Cosmos Studios, 2014.

———. "Sisters of the Sun." *Cosmos: A Spacetime Odyssey*. Studio City, CA: Cosmos Studios, 2014.

———. "A Sky Full of Ghosts." *Cosmos: A Spacetime Odyssey*. Studio City, CA: Cosmos Studios, 2014.

———. "Some of the Things That Molecules Do." *Cosmos: A Spacetime Odyssey*. Studio City, CA: Cosmos Studios, 2014.

———. "Standing Up in the Milky Way." *Cosmos: A Spacetime Odyssey*. Studio City, CA: Cosmos Studios, 2014.

———. "Unafraid of the Dark." *Cosmos: A Spacetime Odyssey*. Studio City, CA: Cosmos Studios, 2014.

———. "When Knowledge Conquered Fear." *Cosmos: A Spacetime Odyssey*. Studio City, CA: Cosmos Studios, 2014.

———. "The World Set Free." *Cosmos: A Spacetime Odyssey*. Studio City, CA: Cosmos Studios, 2014.

Broderick, Damien. *Readng by Starlight*. New York: Routledge, 1995.

Bromley, D Allan. "The Other Frontiers of Science." *Science* 215, no. 4536 (1982): 1035–44.

Brown, William R, and Richard E Crable. "Industry, Mass Magazines, and the Ecology Issue." *The Quarterly Journal of Speech* 59, no. 3 (1973): 259–72.

Brummett. *Rhetoric in Popular Culture*. 4th ed. Thousand Oaks, CA: SAGE, 2015.

Burke, Kenneth. *Language as Symbolic Action*. Berkeley, CA: University of California Press, 1966.

Campbell, Hank. "Cosmos Ratings in the Home Stretch." *Science 2.0*, 2014. http://www.science20.com/science_20/blog/cosmos_ratings_in_the_home_stretch-137670.

Campbell, John Angus. "Charles Darwin and the Crisis of Ecology: A Rhetorical Perspective." *Quarterly Journal of Speech* 60, no. 4 (1974): 442–49.

Campbell, Paul Newell. "The Personae of Scientific Discourse." *Quarterly Journal of Speech* 61, no. 4 (1975): 391–405.

Carmody, John. "Tentative Agreement in Actors' Strike." *The Washington Post*, September 1980. https://www.washingtonpost.com/archive/lifestyle/1980/09/26/tentative-agreement-in-actors-strike/a92eb96b-1848-4d26-af1f-ee5e906a57e8/.

Carter, Bill. "'Civil War' Sets an Audience Record for PBS." *The New York Times*, September 1980. http://www.nytimes.com/1990/09/25/movies/civil-war-sets-an-audience-record-for-pbs.html.

Chamberlain, Charles. "From Haunts to Character: The Meaning of Ethos and Its Relation to Ethics." *Helios* 2, no. 11 (1984): 97–108.

"Close Encounters of the Third Kind." *The Internet Movie Database*, n.d. http://www.imdb.com/title/tt0075860/.

Cole, Dollie. "Public Television." *The Saturday Evening Post*, 1974, 60-61-83.

Cott, Jonathan. "The Cosmos: An Interview with Carl Sagan." *Rolling Stone*, 1980, 43+.

Cragan, John F., and Donald C. Shields. *Applied Communication Research*. Prospect Heights, IL: Waveland, 1981.

"Creationists Demand Airtime on Neil deGrasse Tyson's 'Cosmos'." *The Huffington Post*, March 2014.

Csikszentmihalyi, Mihaly. "Attention and the Holistic Approach to Behavior." In *Flow and the Foundations of Positive Psychology: The Collected Works of Mihaly Csikszentmihalyi*, 1–20. Claremont, CA: Springer, 2014.

———. "Flow with Soul." In *Flow and the Foundations of Positive Psychology: The Collected Works of Mihaly Csikszentmihalyi*, 265–78. Claremont, CA: Springer, 2014.

Cushing, Lincoln. "Posters About Ecology and the Environment Before and During the 1970s." *Electronic Green Journal* 1, no. 30 (2010): 1–5.

Davidson, Keay. *Carl Sagan: A Life*. New York: John Wiley, 1999.

Davis, Harold L. "Needed: More for Public Understanding." *Physics Today* 28, no. 6 (June 1975): 88. doi:10.1063/1.3069041.

Davis, Nicola. "Cosmos: How the Creator of Family Guy Remade Carl Sagan's Pivotal TV Series." *The Guardian*, 2014.

Deez, Darwin. "Constellations." YouTube, 2010.

Doty, William G. "Exploring the Manifold Spheres of Mythos." *Mythosphere* 1, no. 2 (1999): 115–31.

Ebert, Roger. "2001: A Space Odyssey." *RogerEbert.com*, 1998.

Eco, Umberto. *Experiences in Translation*. Edited by Alastair McEwen. Toronto: University of Toronto Press, 2001.

Endres, Danielle, and Samantha Senda-Cook. "Location Matters: The Rhetoric of Place in Protest." *The Quarterly Journal of Speech* 97, no. 3 (2011): 257–82.

Field, Alexander. "Table Dg172-180, Public Broadcasting Stations and Income, by Source: 1961-1997." *Historical Statistics of the United States Millennial Edition Online*. Cambridge University Press, 2015.

Flam, Faye. "What Should It Take to Join Science's Most Exclusive Club?" *Science* 256, no. 5059 (1992): 960–61.

Follette, Marcel Chotkowski La. "Multivariate Perceptions: Reflections on the 147th Annual Meeting of the American Association for the Advancement of Science." *Science* 6, no. 35 (1981): 31–37.

Foss, Sonja K. *Rhetorical Criticism*. Long Grove, IL: Waveland, 2009.

"Fox Networks Group Announces First-Ever Simultaneous Cross-Network Global Premiere Event for 'Cosmos: A Spacetime Odyssey'." *PR Newswire*, 2014.

Gerbner, George. "Science and Scientists on the TV Screen." In *AAAS Annual Meeting*. Toronto, 1981.

Golden, Frederick. "The Cosmic Explorer." *Time*, October 1980, 62+-62+.

Goodell, Rae. "The Visible Scientists." *The Sciences* 17, no. 1 (1977): 6–9. doi:10.1002/j.2326-1951.1977.tb01494.x.

Green, Martin. "Writing About Science." *College English* 43, no. 6 (1981): 569–77. http://www.jstor.org/stable/376994.

Griffin, Em. *A First Look at Communication Theory*. 5th ed. New York: McGraw-Hill, 2003.

Gross, Alan G, and Joseph E Harmon. *Science from Sight to Insight*. Chicago: University of Chicago Press, 2014.

Guincho, El. "Bombay." YouTube, 2010.

Halloran, S Michael. "Aristotle's Concept of Ethos, or If Not His Somebody Else's." *Rhetoric Review* 1, no. 1 (1982): 58–63. http://www.jstor.org/stable/465559.

———. "The Birth of Molecular Biology." *Rhetoric Review* 3, no. 1 (1984): 70–83. http://www.jstor.org/stable/465734.

Hariman, Robert, and John Louis Lucaites. *No Caption Needed*. Chicago: University of Chicago Press, 2007.

Harris, R Allen. "Rhetoric of Science." *College English* 53, no. 3 (1991): 282–307. http://www.jstor.org/stable/378102.

Hyde, Michael J. "Rhetorically, We Dwell." In *The Ethos of Rhetoric*, edited by Michael J Hyde, xiii–xxviii. Columbia, SC: University of South Carolina Press, 2004.

"Inflation Here to Stay?" *The Saturday Evening Post*, April 1974.

Itzkoff, Dave. "'Family Guy' Guy as Astrobiology Guy." *The New York Times*, February 2014.

Jerome, Fred. "On Science and American Society." *Environment* 23, no. 7 (1981): 25–30.

Kevin. "PBS Gets Rated." *Adweek*, December 2009. http://www.adweek.com/tvnewser/pbs-gets-rated/27101.

Kevles, Daniel J. *The Physicists: The History of a Scientific Community in Modern America*. New York: Alfred A. Knopf, 1978.

Kirby, David A. "Science Consultants, Fictional Films, and Scientific Practice." *Social Studies of Science* 33, no. 2 (2003): 231–68.

Kirsch, Jeffrey W. "On a Strategy for Using the Electronic Media to Improve the Public Understanding of Science and Technology." *Science, Technology, and Human Values* 4, no. 27 (1979): 52–58.

Klein, Judy. "The Medium Gets a Message." *Science News* 115, no. 22 (1979): 361+-361+.

Kress, Gunther, and Theo van Leeuwen. *Reading Images: The Grammar of Visual Design*. New York: Routledge, 1996.
Lee, S Young. "Status Report of Public Broadcasting, 1980." Washington, D.C., March 1981. http://files.eric.ed.gov/fulltext/ED205205.pdf.
Leiris, Michel. "The Sacred in Everyday Life." In *The College of Sociology, 1937–39*, edited by Denis Hollier, 24–31. Minneapolis: University of Minnesota Press, 1988.
Lentz, Kirsten Marthe. "Quality versus Relevance: Feminism, Race, and the Politics of the Sign in 1970s Television." *Camera Obscura* 15, no. 1 (2000): 44–93.
Lessl, Thomas M. "Science and the Sacred Cosmos: The Ideological Rhetoric of Carl Sagan." *Quarterly Journal of Speech* 71, no. 2 (1985): 175–87.
———. "The Priestly Voice." *Quarterly Journal of Speech* 75, no. 2 (1989): 183–97.
Lewenstein, Bruce V. "Was There Really a Popular Science 'Boom'?" *Science, Technology, and Human Values* 12, no. 2 (1987): 29–41.
Malone, Adrian. "The Backbone of Night." *Cosmos: A Personal Voyage*. Studio City, CA: Cosmos Studios, 1980.
———. "Blues for a Red Planet." *Cosmos: A Personal Voyage*. Studio City, CA: Cosmos Studios, 1980.
———. "The Edge of Forever." *Cosmos: A Personal Voyage*. Studio City, CA: Cosmos Studios, 1980.
———. "Encyclopedia Galactica." *Cosmos: A Personal Voyage*. Studio City, CA: Cosmos Studios, 1980.
———. "The Harmony of the Worlds." *Cosmos: A Personal Voyage*. Studio City, CA: Cosmos Studios, 1980.
———. "Heaven and Hell." *Cosmos: A Personal Voyage*. Studio City, CA: Cosmos Studios, 1980.
———. "Journeys in Space and Time." *Cosmos: A Personal Voyage*. Studio City, CA: Cosmos Studios, 1980.
———. "The Lives of the Stars." *Cosmos: A Personal Voyage*. Studio City, CA: Cosmos Studios, 1980.
———. "One Voice in the Cosmic Fugue." *Cosmos: A Personal Voyage*. Studio City, CA: Cosmos Studios, 1980.
———. "The Persistence of Memory." *Cosmos: A Personal Voyage*. Studio City, CA: Cosmos Studios, 1980.
———. "Shores of the Cosmic Ocean." *Cosmos: A Personal Voyage*. Studio City, CA: Cosmos Studios, 1980.
———. "Traveller's Tales." *Cosmos: A Personal Voyage*. Studio City, CA: Cosmos Studios, 1980.
———. "Who Speaks for the Earth." *Cosmos: A Personal Voyage*. Studio City, CA: Cosmos Studios, 1980.
Maronie, Sam. "Science to the People." *Starlog*, October 1980, 24+.
McMurran, Kristin. "His Cosmos a Huge Success, Carl Sagan Turns Back to Science and Saturn's Rings." *People*, December 1980. http://www.people.com/people/archive/article/0,,20078104,00.html.
Meister, Mark. "Meteorology and the Rhetoric of Nature's Display." *Quarterly Journal of Speech* 87, no. 4 (2001): 415–28.

"'Mermaids: The New Evidence': Never-Before-Seen Footage of Mermaids Allegedly (VIDEO)." *The Huffington Post*, May 2013. http://www.huffingtonpost.com/2013/05/27/mermaids-the-new-evidence_n_3343773.html.

Merton, Robert K. *The Sociology of Science: Theoretical and Empirical Investigations*. Chicago: University of Chicago Press, 1973.

Miller, Carolyn. "Kairos in the Rhetoric of Science." In *A Rhetoric of Doing: Essays on Written Discourse in Honor of James L. Kinneavy*, edited by Stephen P Witte, Neil Nakadate, and Roger D Cherry, 310–27. Carbondale: Southern Illinois University Press, 1992.

"Mirror Neurons." *Principles of Neural Science*, 2013.

Mittell, Jason. "Narrative Complexity in Contemporary American Television." *Velvet Light Trap* 58 (2006): 29–40.

Nakamura, Jeanne, and Mihaly Csikszentmihalyi. "The Concept of Flow." In *Flow and the Foundations of Positive Psychology: The Collected Works of Mihaly Csikszentmihalyi*, edited by Springer, 239–63. Claremont, CA, 2014.

O'Malley, William J. "Carl Sagan's Gospel of Scientism." *America Magazine*, February 1981. http://americamagazine.org/issue/100/carl-sagans-gospel-scientism.

Orfanos, Spyros D. "Mythos and Logos." *Psychoanalytic Dialogues* 16, no. 4 (2006): 481–99.

Overington, Michael A. "The Scientific Community as Audience: Toward a Rhetorical Analysis of Science." *Philosophy and Rhetoric* 10, no. 3 (1977): 143–64.

Perelman, Chaim, and Lucie Olbrechts-Tyteca. *The New Rhetoric: A Treatise on Argumentation*. Edited by John Wilkinson and Purcell Weaver. London: University of Notre Dame Press, 1969.

Perrault, Sarah Tinker. *Communicating Popular Science*. New York: Palgrave Macmillan, 2013.

Philmus, Robert M. *Visions and Re-Visions: (Re)Constructing Science Fiction*. Liverpool: Liverpool University Press, 2005.

Polanyi, Michael. *Personal Knowledge: Towards a Post-Critical Philosophy*. Harper Tor. New York: Harper & Row, 1964.

Pollack, Gerald A. "The Economic Consequences of the Energy Crisis." *Foreign Affairs* 52, no. 3 (1974): 452–71.

"President Requests More Support for R&D at NSF and NASA." *Physics Today* 31, no. 3 (1978): 101–4.

Preston, Marilynn. "Carl Sagan's 'Cosmos' Will Mean Worlds to You." *The Chicago Tribune*, September 1980.

———. "Looking for Miracles: Seton's Life, Sagan's Last Stand." *The Chicago Tribune*, December 1980.

Reuss, Henry S. "State of the Economy: Unemployment and Inflation." *Vital Speeches of the Day* 40, no. 10 (1974): 295–97.

Rose, Philip. "Philosophy, Myth, and the 'Significance' of Speculative Thought." *Metaphilosophy* 38, no. 5 (2007): 632–53.

Rovin, Jeff. "Carl Sagan's Cosmos." *OMNI*, September 1980, 104–7.

"Science on the Air." *Science News* 118, no. 8 (1980): 115.

Shamos, Morris H. *The Myth of Scientific Literacy*. New Brunswick, NJ: Rutgers University Press, 1995.

Shermer, Michael B. "The View of Science: Stephen Jay Gould as Historian of Science and Scientific Historian, Popular Scientist, and Scientific Popularizer." *Social Studies of Science* 32, no. 4 (2002): 489–524. http://www.jstor.org/stable/3183085.

Stapledon, Olaf. *Star Maker*. Mineola, NY: Dover, 2008.

Steel, Duncan. "Carl Sagan: Practitioner, Popularizer and Proponent of Science." *Contemporary Physics* 42, no. 4 (2001): 247–49.

Sullivan, Dale. "Exclusionary Epideictic: NOVA's Narrative of Excommunication of Fleishmann and Pons." *Science, Technology, & Human Values* 19, no. 3 (1994): 283–306. http://www.jstor.org/stable/690026.

———. "The Ethos of the Epideictic Encounter." *Philosophy and Rhetoric* 26, no. 2 (1993): 113–33.

Sullivan, Dale L. "Kairos and the Rhetoric of Belief." *Quarterly Journal of Speech* 78 (1992): 317–32.

"Terminator 2: Judgement Day." *The Internet Movie Database*, n.d. http://www.imdb.com/title/tt0103064/.

"The Economy : 'Scary' Inflation Rages On, and Everybody Is Getting Angry." *Time* 115, no. 11 (March 1980).

"The Godfather." *The Internet Movie Database*, n.d. http://www.imdb.com/title/tt0068646/.

"The Godfather: Part II." *The Internet Movie Database*, n.d. http://www.imdb.com/title/tt0071562/?ref_=nv_sr_2.

"The Godfather: Part III." *The Internet Movie Database*, n.d.

"The Terminator." *The Internet Movie Database*, n.d. http://www.imdb.com/title/tt0088247/.

Thimmesch, Nick. "The King Kong of Federal Waste." *The Saturday Evening Post*, January 1977, 66+.

Tietge, David J. *Rational Rhetoric: The Role of Science in Popular Discourse*. West Lafayette, IN: Parlor Press, 2008.

"TV Test Tube: PBS Brews Two Science Series." *BioScience* 30, no. 10 (1980): 652.

"Two Science Series Developed for TV." *Science News* 114, no. 16 (1978): 263.

Vatz, Richard E. "The Myth of the Rhetorical Situation." *Philosophy and Rhetoric* 16, no. 3 (1973): 154–61.

Vivante, Paolo. *The Homeric Imagination: A Study of Homer's Poetic Perception of Reality*. Bloomington, IN: Indiana University Press, 1970.

Wallace, Karl R. *Francis Bacon and the Nature of Man*. Urbana: University of Illinois, 1967.

———. *Francis Bacon on Communication and Rhetoric*. Chapel Hill, NC: University of North Carolina Press, 1943.

Walsh, Lynda. *Scientists as Prophets*. New York: Oxford University Press, 2013.

Wander, Philip C. "The Rhetoric of Science." *Western Speech Communication* 40, no. 4 (1976): 226–35.

Wander, Philip C, and Dennis Jaehne. "Prospects for a 'Rhetoric of Science'" *Social Epistemology* 14, no. 2/3 (2000): 211–33.

Weimer, Walter B. "Science as a Rhetorical Transaction: Toward a Nonjustificational Concept of Rhetoric." *Philosophy and Rhetoric* 10, no. 1 (1977): 1–29.

Wenner, Lawrence. “Citizen Involvement in Public Television,” November 1975. http://files.eric.ed.gov/fulltext/ED123676.pdf.
Wilkins, Alasdair. “Ten Tropes You’ll Find in Science Fiction Over and Over Again.” *io9*, September 2010.
Wolf-Chase, Grace. “Coloring the Cosmos.” *Phi Kappa Phi Forum*, 2011, 4–7.
Yoos, George. “A Revision of the Concept of Ethical Appeal.” *Philosophy and Rhetoric* 12, no. 1 (1979): 41–58.
Zakariya, Nasser. “Exhibiting Cosmos.” *Technology and Culture* 56, no. 3 (July 2015): 738–44.
Zender, Mike, and Keith A Crutcher. “Visible Language for the Expression of Scientific Concepts.” *Visible Language* 41, no. 1 (2007): 23–49.

Index

About the Author

Dr. Karen Schroeder Sorensen is assistant professor and coordinator of the Writing Across the Curriculum program at Winona State University in Winona, Minnesota. Originally from Huron, South Dakota, she lived for several years in the Fargo, ND, area. There she earned her PhD in rhetoric, writing, and culture from North Dakota State University. Her research interests currently focus on the rhetoric of popular science and 1970s-era popular culture. After three years in Butte, Montana, she and her family are happily resettled in Minnesota.

www.ingramcontent.com/pod-product-compliance
Ingram Content Group UK Ltd.
Pitfield, Milton Keynes, MK11 3LW, UK
UKHW041123071026
15587UKWH00009B/114

9 781498 507592